Esther Sophie Bischoff

Time-Extended Multi-Robot Task Allocation

A Reoptimization Framework with Provable Performance

Karlsruher Beiträge zur
Regelungs- und Steuerungstechnik
Karlsruher Institut für Technologie

Band 26

Time-Extended Multi-Robot Task Allocation

A Reoptimization Framework with Provable Performance

by
Esther Sophie Bischoff

KIT Scientific Publishing

Karlsruher Institut für Technologie
Institut für Regelungs- und Steuerungssysteme

Time-Extended Multi-Robot Task Allocation - A Reoptimization Framework with Provable Performance

Zur Erlangung des akademischen Grades einer Doktorin der Ingenieurwissenschaften (Dr.-Ing.) von der KIT-Fakultät für Elektrotechnik und Informationstechnik des Karlsruher Instituts für Technologie (KIT) genehmigte Dissertation

von Esther Sophie Bischoff, M.Sc.

Tag der mündlichen Prüfung: 18. März 2025
Hauptreferent: Prof. Dr.-Ing. Sören Hohmann
Korreferent: Prof. Dr. Stefan Nickel

Impressum

KIT Scientific Publishing

Karlsruher Institut für Technologie (KIT)
KIT Scientific Publishing
Straße am Forum 2
D-76131 Karlsruhe

KIT Scientific Publishing is a registered trademark
of Karlsruhe Institute of Technology.
Reprint using the book cover is not allowed.

www.bibliothek.kit.edu/ksp.php | E-Mail: info@ksp.kit.edu | Shop: www.ksp.kit.edu

Print on Demand 2025 – Gedruckt auf FSC-zertifiziertem Papier

ISSN 2511-6312
ISBN 978-3-7315-1449-7
DOI 10.5445/KSP/1000183488

Preface

This thesis is the result of my work as a research assistant at the Institute of Control Systems (IRS) at the Karlsruhe Institute of Technology (KIT). Without the support of many people, this thesis would not have been possible. First and foremost, I would like to thank Prof. Dr.-Ing. Sören Hohmann for offering me this research opportunity. I especially value the fruitful and interesting discussions we had during the past years, your support and the trust you put in me. Furthermore, I would like to thank Prof. Dr. Stefan Nickel for the interest in my work and for the assessment of this thesis. I enjoyed our conversations and appreciate your participation in the evaluation commitee of my thesis defense.

Many thanks to all IRS employees for the very pleasant and cooperative working atmosphere. Especially, I would like to thank Christian and Florian for your support and for many scientific and non-scientific discussions. I really enjoyed sharing a room with you. Moreover, I thank the members of my research group for their support during my research, especially Christian, Jairo, Julian and Sean for proof-reading this thesis and for providing valuable feedback. I am also grateful to all students I supervised during their bachelor or master thesis and who supported my work with their own thesis projects.

Thanks a lot also to my friends for your support and for the great times we spent together during the past years. A huge thanks also to my family. Mama and Papa, thank you for your unconditional love and support. Simon, I am very grateful for your encouragement, support and honesty, which I can always rely on. And Malte, you not only bring us joy and happiness, but sometimes also an important change in perspective. I love you all.

Leonberg, March 2025

Courage doesn't always roar. Sometimes courage is the little voice at the end of the day saying, 'I will try again tomorrow.'

Mary Anne Radmacher

Contents

List of Figures

List of Tables

Abbreviations and Symbols

Abbreviations

Abbreviation	Description
ACO	ant colony optimization
BCnP	branch-and-cut-and-price
BnP	branch-and-price
CBBA	consensus-based bundle algorithm
CBBA-PR	consensus-based bundle algorithm with partial replanning
CD	complex dependencies
CI	cheapest insertion heuristic
CMI	cheapest maximum insertion cost heuristic
CSG	constraint-solution graph
CVRP	capacitated vehicle routing problem
DIH	delete-insert heuristic
DVRP	dynamic vehicle routing problem
eCMI	extended cheapest maximum insertion cost heuristic
eDIH	extended delete-insert heuristic
e. g.	exempli gratia (Latin), translates to "for example"
EGT	evolutionary game theory
et al.	et alii/aliae/alia (Latin), translates to "and others"
GA	genetic algorithm
HCCSP	home care crew scheduling problem
Het.	heterogeneous team of agents
IA	instantaneous assignment
ID	in-schedule dependencies
i. e.	id est (Latin), translates to "this means"/"in other words"
ILP	integer linear program
INI	initial solution approach
LNS	large neighborhood search
MA	memetic algorithm
MD	multiple depot
HPP	hamiltonian path problem
MR	multi-robot tasks
MRS	multi-robot system
MRTA	multi-robot task allocation

Abbreviation	Description
MT	multi-task robots
MTSP	multiple traveling salesman problem
ND	no dependencies
Prec.	precedence constraints
PSO	particle swarm optimization
SA	simulated annealing
SD	standard deviation
SR	single-robot tasks
ST	single-task robots
s. t.	subject to
Sync.	synchronization constraints
TA	time-extended assignment
TDH	task deletion heuristic
TDRP	truck-drone routing problem
TRSP	technician routing and scheduling problem
TS	tabu search
TSP	traveling salesman problem
UAV	unmanned aerial vehicle
UV	unmanned vehicle
VNS	variable neighborhood search
VRP	vehicle routing problem
w. r. t.	with respect to
XD	cross-schedule dependencies

Latin Letters

Symbol	Description
$\mathbf{A}$	assignment matrix
$\tilde{A}$	set of arcs of a CSG
$\tilde{A}_{\mathcal{I},X_{\mathcal{I}}}$	set of arcs of a CSG belonging to problem instance $\mathcal{I}$ and corresponding routing $X_{\mathcal{I}}$ (contains $\tilde{A}_{X_{\mathcal{I}}}$ and $\tilde{A}_{\mathcal{P}}$)
$\tilde{A}_{\mathcal{P}}$	set of precedence arcs of a CSG
$\tilde{A}_{X_{\mathcal{I}}}$	set of routing arcs of a CSG corresponding to routing $X_{\mathcal{I}}$
a	capability
a_i^m	capability of agent with iterate m to execute task with iterate i
$a_{N+1}^{\max}$	maximum of all agents' capabilities to perform the inserted task n_{N+1}
$a_{N+1}^{\min}$	minimum of all agents' capabilities to perform the inserted task n_{N+1}
b_{ir}^m	binary variable in the BnP approach indicating whether vertex with iterate i is part of the route of agent with iterate m
b_{ijr}^m	binary variable in the BnP approach indicating whether vertices with indices i and j are part of the route of agent with index m with i being scheduled before j
c^e	part of the objective function value associated with task execution times
c^t	part of the objective function value associated with transitioning times
c^w	part of the objective function value associated with waiting times
c_{dyn}	constant weighting the dynamic penalty component within the GA
c_{stat}	static penalty component within the GA
d_i^m	duration needed by agent with iterate m to execute task with iterate i
$d_{i,j}^m$	duration needed by agent with iterate m to transition between two vertices specified by iterates i and j
E	set of edges
f	fitness function applied in the GA
G	graph representing a time-extended MRTA problem instance
$\tilde{G}$	constraint-solution graph (CSG)
$\tilde{G}_{\mathcal{I},X_{\mathcal{I}}}$	CSG of problem instance $\mathcal{I}$ and corresponding routing $X_{\mathcal{I}}$
$g_{\mathcal{C}}$	cluster weight of cluster $\mathcal{C}$ applied in the BnP cluster method
g_i	task weight of the task with iterate i applied in the BnP cluster method
i	iterate used for tasks or vertices (i. e. tasks and depots)
i^*	index of the vertex precedent to a specific task within a globally optimal solution
J	objective function
$J_{\mathcal{I}}^*$	objective function value of a globally optimal solution to problem instance $\mathcal{I}$
j	iterate used for tasks or vertices (i. e. tasks and depots)
j^*	index of the vertex following a specific task within a globally optimal solution
K	total number of agents in the set of agents $\mathcal{K}$

Symbol	Description
k	agent
k_m	agent specified by iterate m
l	running index, iteration number
lev	function determining the Levenshtein distance between two routings
m	iterate used for agents
m^*	iterate of the agent a specific task is assigned to in a globally optimal solution
N	total number of tasks in the set of tasks $\mathcal{N}$
n	task
n_i	task specified by iterate i
$n_{L3.2.2}$	number of cycles within a CSG
n_{N+1}	additional task corresponding to the modification of task insertion
n_ω	number of direct subnodes of an operator ω in the BnP decision variable method that must be fulfilled in order for ω to be filfulled
o	depot
o_m	depot of the agent specified By iterate m
$o_{\mathcal{K}}$	common depot of all agents
P_{het}	problem configuration considering heterogeneous agents and no precedence or synchronization constraints
P_{hom}	problem configuration considering homogeneous agents and no precedence or synchronization constraints
$P_{\mathcal{P},\mathcal{S},\text{het}}$	problem configuration considering heterogeneous agents, precedence and synchronization constraints
$p_{i,j}$	precedence constraint between tasks with iterates i and j
$p^{+}_{\tilde{i},\tilde{j}}$	inserted precedence constraint
$p^{-}_{\tilde{i},\tilde{j}}$	deleted precedence constraint
q	index of the modified task for the modifications of task deletion, task position variation, and task duration variation, or of the modified agent for the modifications of agent capability variation and agent velocity variation
r	route in the BnP approach
$\mathbf{S}$	scheduling matrix
$s_{i,j}$	synchronization constraint between tasks with iterates i and j
$s^{+}_{\tilde{i},\tilde{j}}$	inserted synchronization constraint
$s^{-}_{\tilde{i},\tilde{j}}$	deleted synchronization constraint
$s^{\mathcal{N}}_{i\omega}$	binary variable in the BnP decision variable method indicating whether the simple task with index i is a direct subnode of operator ω
$s^{\Omega}_{\tilde{\omega}\omega}$	binary variable in the BnP decision variable method indicating whether operator $\tilde{\omega}$ is a direct subnode of operator ω
$T_{\max}$	large constant
$t_\square$	start time of execution of task with iterate $\square$ or time at which an agent leaves its depot $\square$
V	set of vertices

Symbol	Description
$\tilde{V}$	set of synchronized vertices of a CSG
$\tilde{V}_{\mathcal{I}}$	set of synchronized vertices of a CSG belonging to problem instance $\mathcal{I}$
v	velocity of an agent
$\tilde{v}$	synchronized vertex
v_m	velocity of the agent specified by iterate m
$v^{\max}$	maximum of the velocities of all agents
$v^{\min}$	minimum of the velocities of all agents
$w_{i,j}$	waiting time between vertices with indices i and j
X	routing information
$X_{\mathcal{I}}$	routing information corresponding to problem instance $\mathcal{I}$
$\vec{X}$	solution to an MRTA problem instance
$\vec{X}_{\mathcal{I}}$	solution to MRTA problem instance $\mathcal{I}$
$\vec{X}_{\mathcal{I}}^{*}$	globally optimal solution to MRTA problem instance $\mathcal{I}$
$\vec{X}_{\mathcal{I}}^{\text{aprx}}$	approximate solution to MRTA problem instance $\mathcal{I}$
$\vec{X}_{\mathcal{I}}^{\text{reo}}$	reoptimized solution to MRTA problem instance $\mathcal{I}$
$\vec{X}_{\mathcal{I}_{\text{mod}}}^{\text{reo,GA}}$	reoptimized solution to MRTA problem instance $\mathcal{I}$ generated by the GA-based reoptimization framework
$\vec{X}_{\mathcal{I}_{\text{mod}}}^{\text{reo,heur}}$	reoptimized solution to MRTA problem instance $\mathcal{I}$ generated by a reoptimization heuristic
$\vec{X}_{\mathcal{I}_{\text{mod}}}^{\text{reo,meta}}$	reoptimized solution to MRTA problem instance $\mathcal{I}$ generated by a meta-heuristic reoptimization framework
$x_{i,j}^{m}$	binary routing variable indicating whether edge (i,j) lies on the route of agent m

Greek Letters

Symbol	Description
α	approximation ratio
β_i	maximum number of agents' routes affected by a temporal shift of vertex with iterate i
Δ^{+}	increment in the objective function value resulting from solving a MRTA task insertion reoptimization problem instance
Δ^{-}	difference in the objective function value resulting from solving a MRTA task deletion reoptimization problem instance
Δ_{c^e}	difference in task execution times between two solutions
Δ_{c^t}	difference in transition times between two solutions
Δ_{c^w}	difference in waiting times between two solutions
$\Delta^{i,j,m}$	insertion costs of inserting a task between the vertices with indices i and j on the route of the agent with iterate m

Symbol	Description
$\Delta_{\max}^{i,j,m}$	maximum insertion costs of inserting a task between the vertices with indices i and j on the route of the agent with iterate m
$\Delta_{\mathrm{emax}}^{i,j,m}$	extended maximum insertion costs of inserting a task between the vertices with indices i and j on the route of the agent with iterate m
Δ^*	optimal insertion costs for problems of configuration P_{het} or P_{hom}
$\Delta_{\max}^*$	optimal maximum insertion costs
Δ_{emax}^*	optimal extended maximum insertion costs
Δ^{TDH}	difference in the objective function value resulting from applying the TDH reoptimization heuristic
δ	distance
$\delta(i,j)$	distance between the two vertices with indices i and j
ϵ	parameter weighting the influence of waiting times on the objective function
γ	parameter weighting the influence of task execution times on the objective function
κ	parameter applied in the mutation operator of the GA; cost associated with a route in the BnP approach
λ_r^m	binary variable indicating whether route r is chosen in the BnP approach
μ_ω	binary decision variable in the BnP decision variable method indicating whether operator ω is fulfilled
Ω	set of operators in the BnP decision variable method
ω	`AND` or `OR` operator in the BnP decision variable method
$\phi_{\mathcal{K}\setminus m}$	sum of objective function value associated with all agents in $\mathcal{K}$ other than m
ϕ_m	objective function value associated with the route of agent with iterate m
$\varphi_m(i,j)$	objective function value associated with the route of agent with iterate m without the part between the vertices with indices i and j
τ	basic task duration
τ_i	basic duration of the task specified by iterate i
ζ	delay in the BnP approach
ζ_{ij}	delay before the vertex with iterate j caused by a delay of the vertex with iterate i which are both in the route of the same agent
ζ_i^m	delay right before the vertex with iterate i in the route of the agent with iterate m

Calligraphic and Other Symbols

Symbol	Description
$\mathbb{N}$	set of natural numbers

Symbol	Description
$\mathbb{R}$	set of real numbers
$\mathcal{A}$	set of agents' capabilities
$\mathcal{B}_{\mathcal{C}}$	set of possible decompositions of a cluster $\mathcal{C}$ in the BnP cluster method
$\mathcal{B}_{\mathcal{C}}^{\text{inv}}$	set of decompositions of a cluster $\mathcal{C}$ that do not form a valid minimal decomposition of $\mathcal{C}$ in the BnP cluster method
$\mathcal{B}_{\mathcal{C}}^{\text{val}}$	set of valid minimal decompositions of a cluster $\mathcal{C}$ in the BnP cluster method
$\mathcal{C}$	cluster in the BnP cluster method
$\mathcal{C}_{\Sigma}$	set of all clusters in the BnP cluster method
$\mathcal{D}$	set of distances between nodes
$\mathcal{E}$	set of direct subtasks of an operator in the BnP decision variable method
$\mathcal{I}$	(initial) MRTA problem instance
$\mathcal{I}_{\text{mod}}$	modified MRTA problem instance
$\mathcal{I}^{+}$	modified problem instance corresponding to the modification of task insertion
$\mathcal{I}^{-}$	modified problem instance corresponding to the modification of task deletion
$\mathcal{I}^{a\updownarrow}$	modified problem instance corresponding to the modification of agent capability variation
$\mathcal{I}^{\delta\updownarrow}$	modified problem instance corresponding to the modification of task position variation
$\mathcal{I}^{\text{p+}}$	modified problem instance corresponding to the modification of precedence constraint insertion
$\mathcal{I}^{\text{p-}}$	modified problem instance corresponding to the modification of precedence constraint deletion
$\mathcal{I}^{\text{s+}}$	modified problem instance corresponding to the modification of synchronization constraint insertion
$\mathcal{I}^{\text{s-}}$	modified problem instance corresponding to the modification of synchronization constraint deletion
$\mathcal{I}^{\tau\updownarrow}$	modified problem instance corresponding to the modification of task duration variation
$\mathcal{I}^{v\updownarrow}$	modified problem instance corresponding to the modification of agent velocity variation
$\mathcal{K}$	set of agents
$\mathcal{L}$	set of direct suboperators of an operator in the BnP decision variable method
$\mathcal{N}$	set of tasks
$\mathcal{N}_k$	set of tasks allocated to agent k
$\mathcal{O}$	set of depots
$\mathcal{P}$	set of precedence constraints
$\mathcal{R}$	set of routes in the BnP approach
$\mathcal{S}$	set of synchronization constraints

Symbol	Description
$\mathcal{T}$	set of tasks' basic durations
$\mathcal{V}$	set of agents' velocities

Indices, Exponents and Operators

Symbol	Description
$\lvert\square\rvert$	number of elements in set $\square$
$\overline{\square}$	velocity or distance $\square$ normalized w. r. t. $v^{\max}$ or capability or task duration $\square$ normalized w. r. t. $a_{N+1}^{\max}$
$\square_{>0}$	set $\square$ containing positive values
$\square_{\geq 0}$	set $\square$ containing non-negative values
$\square^{+}$	set $\square$ belonging to the modification of task insertion
$\square^{-}$	set $\square$ belonging to the modification of task deletion
$\square^{a\updownarrow}$	set $\square$ belonging to the modification of agent capability variation
$\square^{\delta\updownarrow}$	set $\square$ belonging to the modification of task position variation
$\square^{\text{p+}}$	set $\square$ belonging to the modification of precedence constraint insertion
$\square^{\text{p-}}$	set $\square$ belonging to the modification of precedence constraint deletion
$\square^{\text{s+}}$	set $\square$ belonging to the modification of synchronization constraint insertion
$\square^{\text{s-}}$	set $\square$ belonging to the modification of synchronization constraint deletion
$\square^{\tau\updownarrow}$	set $\square$ belonging to the modification of task duration variation
$\square^{v\updownarrow}$	set $\square$ belonging to the modification of agent velocity variation
$\square_{\mathcal{I}}^{\text{CMI}}$	routing, solution or corresponding objective function value $\square$ generated by the application of the CMI reoptimization heuristic to solve problem instance $\mathcal{I}$
$\square_{\mathcal{I}}^{\text{DIH}}$	routing, solution or corresponding objective function value $\square$ generated by the application of the DIH reoptimization heuristic to solve problem instance $\mathcal{I}$
$\square_{\mathcal{I}}^{\text{eDIH}}$	routing, solution or corresponding objective function value $\square$ generated by the application of the eDIH reoptimization heuristic to solve problem instance $\mathcal{I}$
$\square_{\mathcal{I}}^{\text{INI}}$	routing, solution or corresponding objective function value $\square$ generated by the application of the INI reoptimization heuristic to solve problem instance $\mathcal{I}$
$\square_{\mathcal{I}}^{\text{TDH}}$	routing, solution or corresponding objective function value $\square$ generated by the application of the TDH reoptimization heuristic to solve problem instance $\mathcal{I}$

1 Introduction

Over the past years, the interest in *multi-robot systems (MRSs)* has grown constantly [YJC13, SKRJ20, RAT19]. One major cause is the potential improvement in standard of living associated with their deployment [Man13, BKAJ15, BFC$^+$16, DBW$^+$18]. Supporting that, robotic capabilities increase rapidly [Oxf] and especially connected and cooperating robotic systems allow for accomplishing complex tasks in various domains [RAT19].

In MRSs, a group of robots, that can either be homogeneous or heterogeneous, works together towards a common goal [KHE15]. Compared to individual robots, MRSs have several advantages [GM12, YJC13, PRS16, VR21]: They can achieve better overall system performance, e. g. due to task parallelization. Furthermore, they potentially are more reliable, flexible, versatile, robust, and obviously allow for a better spatial distribution than a single robot. Due to these promising advantages, the areas of application for MRSs are manifold. They include planetary exploration [SMB$^+$20, WMS$^+$21, WMS$^+$22], e-commerce and automated warehouses [KYCL18, YZM$^+$21, WC21], automated production and assembly [AAM13, DDD$^+$17, GWS18, BPS$^+$20], exploration and mapping [RDW$^+$00], agriculture [CFSR23], patrolling [FIN17], search and rescue [HYBB20], and inspection scenarios for the onshore oil and gas industry [SK16a] as well as in solar [MSF$^+$14] and thermosolar power plants [MFGC21].

To successfully deploy MRSs, the problem of *multi-robot task allocation* (MRTA) must be solved [GM04]. MRTA plays an important role in multi-robot coordination, which is essential for the application of MRSs since it highly influences the overall system performance [YJC13, VR21]. MRTA answers the question of *task allocation* that determines which tasks are performed by which robots. Depending on the application setting, also *task scheduling*, i. e. the determination of the order of tasks, and *task decomposition*, i. e. the choice of a set of subtasks to perform a task with different possible realizations, are to be answered by MRTA. [GM04, KSD13, ASGM23]

However, MRTA problems are generally complex to solve. All MRTA problems that consider both task allocation and task scheduling are strongly NP-hard [GM04]. Due to the complexity of MRTA, its numerous fields of application with different kinds of constraints, and its importance for the overall MRS performance, various MRTA approaches have already been proposed [SKRJ20].

Nevertheless, these automated MRTA algorithms can only generate solutions that are meaningful and thus purposeful in praxis, if the model of the problem to solve represents all relevant aspects of the real world problem under consideration [MKF$^+$15]. An expressive model may for example consist of a definition of all tasks and robotic systems including their relevant characteristics such as, among many others, the robots' capabilities to perform each task and constraints on some tasks' orders. While a meaningful automated problem modeling may be possible in structured environments, e. g. in an automated warehouse with clearly defined procedures, generating an expressive model in unstructured environments is a challenging task. It requires situational awareness and recognition as well as the capability of abstraction and strategic thinking, all of which are skills in which humans outclass computers [CSP+19, Rot22].

1.1 Towards Interactive Optimization in Multi-Robot Task Allocation

Combining the computational strengths of the automation with the cognitive abilities of the human by pursuing an interactive MRTA approach promises high synergy potential. In the current field of research in the application of robotic systems for space exploration for example, the robotic mission must be conducted in an a priori rather unknown and unstructured environment. In this example, the tasks to be performed by the robotic team are investigations that may include the mapping of regions of interest, taking images of various kinds, or the sampling of soil [WMS+21, WMS+22]. Here, it is the state of the art to let experienced scientists define both the locations and the type of investigations to be performed there [WMS+21, WMS+22]. To solve MRTA problems of this kind both task allocation as well as task scheduling must be determined, i. e. which tasks are to be performed by which robot and in which order. Another current research area is the development of service robots for application in nursing practice [Hol, LV]. In this context, robots can be applied to provide care utensils or food, automatically control material stocks or to support patients while walking [Hol, LV]. Also in such an MRS application, the capturing of all relevant tasks and constraints, e. g. on tasks' orders, in a dynamic hospital environment can be done best by a human operator who knows about current patients, priorities, workloads and procedures.

While letting humans model the MRTA problem and using algorithms and computers to solve it is a good attempt to generate synergies, modifications to the model of the problem instance may often be necessary. One reason for this may be if the human operator realizes that the MRTA solution generated violates some real-life constraints that were forgotten in the initial model. For example, in the application of service robots in nursing practice, the operator may want a robot to accompany a patient to some check-up prior to the patient being served food. If an MRTA solution violates this precedence constraint, it will need to be added to the problem model, i. e. the problem is modified by a precedence constraint insertion. Besides precedence constraints also synchonization constraints forcing the simultaneous execution of some tasks may be added, e. g. food and medicaments being served at the same time to a patient. This corresponds to the modification of synchronization constraint insertion. Also in the case the human operator realizes that an additional task should be performed, a modification to the problem instance in the form of a task insertion is necessary. The scientist might for example recognize another interesting area and thus add an investigation task to the space exploration mission. If the scientist wants to change the position where an investigation should take place, this corresponds to the modification of task position variation. Moreover, if the human operator considers one of the previously defined tasks not to be needed, the corresponding task deletion modifies the problem instance. In the nursing scenario for example, a patient might have been moved to another ward and thus not need to be served food by the service robots on the old ward. More possible problem modifications include the variation of a robot's velocity or of its capability to perform certain tasks, e. g. after a software or hardware update, the variation of a task's duration as well as the deletion of unnecessary precedence or synchronization constraints. In these occasions, in order to solve the real-world problem under consideration and due to the MRTA solution's relevance for the overall system performance [YJC13, VR21], good solutions to the modified MRTA problem instances are required quickly. This allows the operator to evaluate the new solution in time and to potentially make further adjustments, if required.

Consequently, not only is it promising to combine the strengths of human and automation by letting the human initially define the problem instance which is subsequently solved by automated algorithms, but also should it be possible to allow for modifications of the problem instance. This approach of the human refining the problem instance through modifications to the problem model which is then in turn solved anew by the automation, yields an *interactive optimization system* with a *problem-oriented interaction* [MKF$^+$15]. In the context of MRTA problems, such a system is called *interactive MRTA optimization system* in this thesis. A schematic representation of such an interactive MRTA optimization system is depicted in Figure 1.1. Not only does this approach tackle the problem of generating expressive models of the real-world problems under consideration through the incorporation of human expertise, but also does an active human involvement increase human trust in the automation generating the solution to the coordination problem [MKF$^+$15].

When it comes to solving the modified problem instances, especially in direct interaction with a human operator, several aspects are relevant:

- *High quality solutions* are essential for the overall MRS performance [YJC13]. A high quality solution implies optimizing the given objective function while respecting all constraints in order for the solutions to be executable in praxis. Especially due to the importance of MRTA solutions for the overall MRS performance [GM04, VR21], *performance guarantees* for the applied MRTA solution approaches are desirable [Gin17].

- Additionally, a fast incorporation of the modifications is necessary for user acceptance of the system [HGQ$^+$12, MKF$^+$15]. This aspect, referred to as *responsiveness* [HGQ$^+$12], describes the computation time needed to generate a new solution after the modification to the optimization problem has been made.

- Another aspect that is especially important in an interactive setting is the recognizability of the initial solution. Since human problem modifications are often

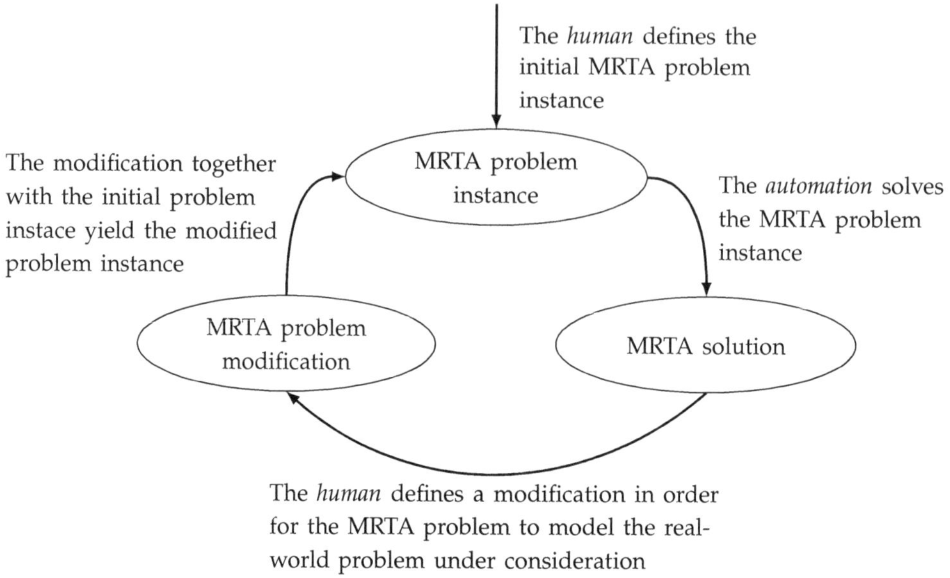

Figure 1.1: Schematic representation of an interactive MRTA optimization system.

directly related to the initial solution, computing a solution that is as close as possible to the initial one is important for user acceptance [HGQ+12, MKF+15]. This aspect is referred to as *solution stability* [HGQ+12].

Additionally, also in settings without direct user interaction, the necessity to react to modifications to MRTA problem instances occurs. In an automated production hall for example, the orders to process may either be defined by a human operator or they might be processed automatically. In both cases, the cancellation of an order requires modifications to the MRTA problem instance. Also in a scenario of robots inspecting e. g. power plants, automatically activated alerts may cause additional inspections to become necessary and thus also modify the MRTA problem instance. Also in such settings of automatically generated modifications to the problem instance, it is essential to generate suitable high quality solutions quickly. The criteria of responsiveness and high quality solutions thus remain significant, independently of whether modifications to the MRTA problem instance are generated automatically or by a human operator.

Existing MRTA solution approaches mainly consist of centralized optimization-based approaches and decentralized market-based approaches [SKRJ20, APP+22]. In the vast majority of market-based approaches, several robots evaluate locally available information to generate a solution to the MRTA coordination problem. However, the distribution of knowledge on the features of the overall coordination problem restrains the capability of decentralized approaches to globally optimize the common objective [APP+22].

In the interactive setting, where the model of the problem instance is defined and refined by a human operator and subsequently solved by an automated MRTA algorithm, it is a valid assumption that the global information of the model of the MRTA problem under consideration is available to a central coordination entity. Thus, the human interacts and defines the model of the problem instance for a central computation unit which generates the solution to the respective instance. Once the solution is finalized and approved by the human operator, it will be communicated to the robots to execute the solution. Having access to global information is what most optimization-based methods for MRTA problems rely on in order to solve the coordination problem [APP+22]. They search a solution space to find solutions either minimizing or maximizing some optimization objective while respecting given problem related constraints. Optimization-based approaches include both exact and approximative methods [SKRJ20, APP+22]. While exact approaches solve problem instances to optimality and thus perfectly fulfill the requirement of producing high quality solutions, they generally do not meet the criterion on responsiveness due to the computational complexity of MRTA coordination problems [APP+22]. Approximative methods on the other hand, such as heuristics and metaheuristics, are computationally much more efficient than exact approaches [APP+22], but typically lack performance guarantees [Gin17]. Furthermore, the majority of optimization-based approaches solves each problem instance independently. Thus, when solving a modified problem instance, the stability of the solution is generally not considered, since the modified problem instance is solved without consideration of the previous solution to the initial instance.

In contrary to solving every problem instance individually via optimization, the concept of *reoptimization* is based on the idea of utilizing the knowledge of an optimal solution to an optimization problem when solving a slightly modified problem instance [BHMW08, ABE09]. When applied to NP-hard optimization problems, reoptimization aims at finding an approximate solution to the modified instance of better quality than what would have been possible without the knowledge of the initial solution, and/or generating an approximate solution of at least the same quality but in a computationally more efficient manner [BHMW08, ABE09]. Thus, reoptimization is a promising approach when it comes to finding a compromise between generating high quality solutions in an at the same time responsive interactive MRTA optimization system. This holds especially if performance guarantees w. r. t. the solution quality can be given. Furthermore, if the reoptimization approach only slightly modifies the initial solution, it is likely to increase the stability of solutions compared to classical optimization methods [BP11].

Reoptimization methods with performance guarantees have already been successfully introduced for several optimization problems such as the steiner tree problem [BBH+08, EMP09], the knapsack problem [ABS10] and the traveling salesman problem (TSP) [ABS03, AEMP09, BFH+06, BK10]. In MRTA with task allocation and task scheduling however, to the best of the author's knowledge so far no reoptimization approaches, especially ones giving performance guarantees, have been proposed.

1.2 Research Contribution

The goal of this thesis is to develop responsive MRTA reoptimization approaches with performance guarantees that furthermore yield stable solutions. This enables interactive optimization systems for MRTA problems which incorporate problem modifications into given MRTA solutions, as described above.

In order to be suitable for the interactive setting and to allow for a variety of applications as set out in Section 1.1, the following requirements and assumptions are imposed on the sought solution approaches:

In the interactive target setting, it is assumed that the model of the problem instance is defined and refined by a human operator and subsequently solved by an automated MRTA algorithm, i. e. the human interacts with a central computation unit. Therefore, it is assumed that all information on problem data is centrally available which is why only **centralized solution approaches** are considered.

Since the information on all robots as well as on all tasks to be performed is available for coordination, tasks must be allocated to robots and additionally the timing information of the tasks, i. e. the tasks' schedule, must be determined to generate expressive solutions. Consequently, an approach that considers **task allocation as well as task scheduling** is required, i. e. a so called **time-extended MRTA problem**[1] is considered.

To allow for a variety of application domains including e. g. the space example outlined above, heterogeneous teams of **robots with different capabilities** to perform a set of **different kinds of tasks** shall be considered. Moreover, robots may have **different transitioning velocities** between tasks. Furthermore, the consideration of temporal constraints between tasks including **precedence and synchronization constraints** shall be possible.[2] Precedence constraints require a precedent task to be finished before the execution of the following task can start, and synchronization constraints require some tasks to start simultaneously. Synchronization constraints also allow for the modeling of cooperative tasks that require direct interaction between robots.

Given these requirements, the following contributions are made within this thesis:

It is of relevance that all solutions generated by the reoptimization methods respect all constraints of the problem instance and thus are executable in praxis. A first contribution of this thesis is the definition of a feasibility criterion for solutions to MRTA problem instances with the properties stated above.

In order to reach the goal of introducing responsive reoptimization approaches with performance guarantees for MRTA problems which are needed for interactive optimization systems as motivated in Section 1.1, the main contributions of this thesis towards this goal are the following:

---

[1] More explanation on MRTA problems with time-extended assignment is given in Section 2.2.

[2] MRTA problems with the considered features are denoted as heterogeneous, time-extended MRTA problems with precedence and synchronization constraints in the following.

1. Reoptimization heuristics with performance guarantees for ten different modifications to heterogeneous, time-extended MRTA problem instances with precedence and synchronization constraints are introduced. All of them guarantee the resulting reoptimized solution to be feasible for the respective modified problem instance. Furthermore, the proposed heuristics for the modifications of adding a task to and deleting a task from such an MRTA problem instance, are the first reoptimization heuristics for MRTA problems for which guarantees on the quality of the resulting solutions are given. Upper bounds on how much the quality of a solution generated by the respective reoptimization heuristic can, in the worst case, differ from a globally optimal solution are proven.
2. A reoptimization framework that combines metaheuristic solution approaches with the previously defined reoptimization heuristics is introduced. Metaheuristic algorithms combine intensification, i. e. the process of improving solution quality within certain limits of the search space by applying slight perturbations to previous solutions, and diversification, i. e. a process that aims at searching large areas of the entire search space [ZBB10, Chapter 4.3] and thus aim at searching wider areas of the search space to potentially further improve the heuristic solution. A specific implementation based on a genetic algorithm (GA) is suggested. The application of the so-called *elitism* mechanism guarantees the resulting solutions to be at least as good as the solutions obtained by the introduced reoptimization heuristics. Hence, the performance guarantees given for the reoptimization heuristics remain valid.
3. An extensive evaluation to analyze the performance of the proposed reoptimization heuristics as well as of the GA-based reoptimization framework compared to an exact solution approach and a conventional GA is conducted. All requirements arising from the intended interactive application, i. e. quality of the solutions, responsiveness of the solution approaches and solution stability are analyzed via appropriate performance measures. The influence of various MRTA problem features on the performance measures is inspected. These features include problem size, considered heterogeneity of the robots and tasks, and number of precedence and synchronization constraints considered. The evaluations yield first evidence of the suitability of the proposed reoptimization approaches for their application in interactive MRTA optimization systems. Especially the reoptimization heuristics provide promising results w. r. t. the combination of the criteria relevant for application in the intended interactive setting. The evaluation also indicates in which cases further solution improvement via the GA-based reoptimization framework is beneficial.

In reoptimization, guarantees on the solution quality can in general only be given if the initial solution is globally optimal. Thus, for the evaluation of the proposed reoptimization heuristics, an exact solution approach is required. In the current state of the art a variety of MRTA solution approaches exist that consider both task allocation and task scheduling and can account for various constraints, as required for the evaluation within this thesis. However, to the best of the author's knowledge, no exact

MRTA solution approach exists that additionally to task allocation and task scheduling also considers the problem of task decomposition. An additional contribution of this thesis is therefore the incorporation of the problem of task decomposition into an exact MRTA solution approach. Three approaches are presented that, for the first time, allow for the consideration of task decomposition additionally to task allocation and task scheduling. The *decomposition method* decouples the problem of task decomposition from task allocation and scheduling, while the *decision variable method* and the *cluster method* incorporate task decomposition explicitly into the overall optimization model. Since task decomposition is not in the focus of this thesis, the presentation of these approaches is given in Appendix B. Also, a brief evaluation to compare the proposed approaches and to prove their ability to solve respective MRTA problems is given.

The resulting structure of the remaining thesis is depicted in Figure 1.2. In Chapter 2, the current state of the art on optimization and reoptimization approaches for MRTA problems is presented. Upon this, a more detailed formulation of the research questions and contributions of this thesis is given. A formal definition of heterogeneous, time-extended MRTA optimization and reoptimization problems with precedence and synchronization constraints and the introduction of a feasibility criterion for respective solutions is given in Chapter 3. This is followed by the introduction and theoretical analysis of reoptimization heuristics for ten different problem modifications. In Chapter 4, the extension of the proposed heuristics towards a metaheuristic reoptimization frameword is introduced. Finally, an extensive evaluation of the proposed reoptimization approaches comparing them to two state of the art optimization methods and investigating the influence of different problem features is presented in Chapter 5.

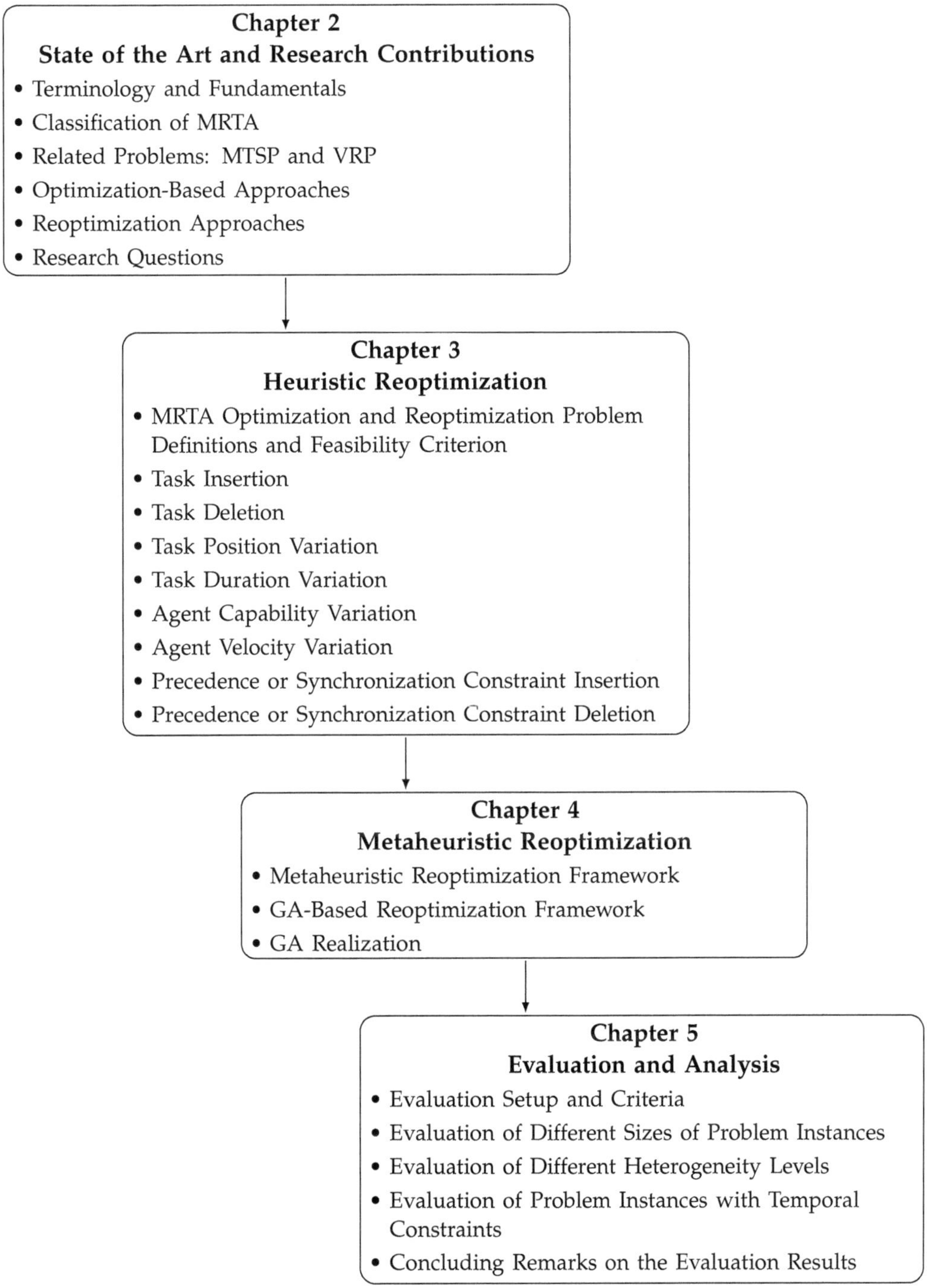

Figure 1.2: Structure of the thesis.

2 State of the Art and Research Questions

In this chapter, an overview of the state of the research on optimization and reoptimization methods in time-extended MRTA problems, i. e. MRTA problems considering both task allocation and scheduling as outlined in Section 1.2, is given. The terminology used throughout the thesis as well as relevant fundamentals are given in Section 2.1 followed by an introduction of the existing MRTA taxonomies in Section 2.2. Problems closely related to the considered class of MRTA problems are presented in Section 2.3. Optimization-based solution approaches for MRTA and related problems are presented in Section 2.4. An overview of existing reoptimization methods to solve both MRTA as well as related problems is given in Section 2.5. Based on this analysis of the state of the art, the research gap and the research questions to be answered in this thesis are formulated in Section 2.6.

2.1 Terminology and Fundamentals

The area of time-extended MRTA and related problems is a wide field with various applications. In many of these domains, different formulations are used which is why a variety of formulations exists. Therefore, the terminology used throughout this thesis is defined in this section. Furthermore, relevant basic concepts are introduced.

Tasks and agents are the two most important terms in order to define MRTA problems. In this thesis, a task is defined as given in Definition 2.1.

> **Definition 2.1 (Task)**
> *A task is a job that can be performed by one or several agents. A task can be associated with a location and may require certain capabilities of the executing agent(s).*

Zlot [Zlo06] introduces different task types suitable for MRTA problems: *Decomposable* tasks can be represented as a set of subtasks out of which some subset of subtasks must be executed. Single jobs to be performed by one agent that are not decomposable into subtasks are denoted as *elemental* or *atomic* tasks. Decomposable tasks for which there exists only one possible way of decomposing them into subtasks are differentiated into *decomposable simple* and *compound* tasks. The subtasks of decomposable simple tasks must all be performed by the same agent, while the subtasks of compound tasks may be allocated to different agents. Tasks that cannot be split to be allocated to several agents are called *simple* tasks. They comprise elemental tasks as well as decomposable simple tasks. If there exist several possibilities how to decompose a task into subtasks and the subtasks of at least one decomposition can be allocated to multiple agents, these tasks are called *complex* tasks. An overview of the relations between the different task types is depicted in Figure 2.1.

Depending on the application domain, tasks can e. g. also be represented by cities or customers in transportation or routing problems or by patients in healthcare applications.

Tasks of all types are to be performed by agents, which are defined as follows.

Definition 2.2 (Agent)
An agent is an autonomous entity that is capable of executing tasks and of transitioning between tasks. Each agent can have individual transition velocities and capabilities to perform different tasks.

If all agents considered in an MRTA problem have identical features, i. e. all agents have identical capabilities for all tasks and identical transition velocities, this group of agents is called *homogeneous*. If the agents differ in at least one of the two features, the group of agents is denoted as *heterogeneous*. Depending on the application domain, agents might e. g. be robots in general robotic applications, vehicles or salesmen in transportation or routing problems or caretakers in healthcare applications.

To define MRTA problems, task allocation, task scheduling, and task decomposition are relevant, which are described in Definitions 2.3, 2.4 and 2.5, respectively. The definitions are similar to the ones given by Antonyshyn el al. [ASGM23].

Definition 2.3 (Task allocation)
Given a set of agents $\mathcal{K}$ and a set of simple tasks $\mathcal{N}$, a task allocation (or agent-task allocation) is an unambiguous assignment of tasks to agents. Each task $n \in \mathcal{N}$ is assigned to one agent $k \in \mathcal{K}$ such that each agent $k \in \mathcal{K}$ is allocated a (potentially empty) subset of tasks $\mathcal{N}_k \subset \mathcal{N}$ out of the overal set of tasks.

MRTA problems always comprise the problem of task allocation [Ger03]. Depending on the specific problem under consideration, also task scheduling is sought, which is defined as follows.

Definition 2.4 (Task scheduling)
Given a set of simple tasks $\mathcal{N}$, task scheduling defines a permutation of the tasks $n \in \mathcal{N}$, i. e. a sequence of task execution.

For task allocation and task scheduling, simple tasks are required. Thus, compound tasks must be represented by their decomposition into simple subtasks. If complex

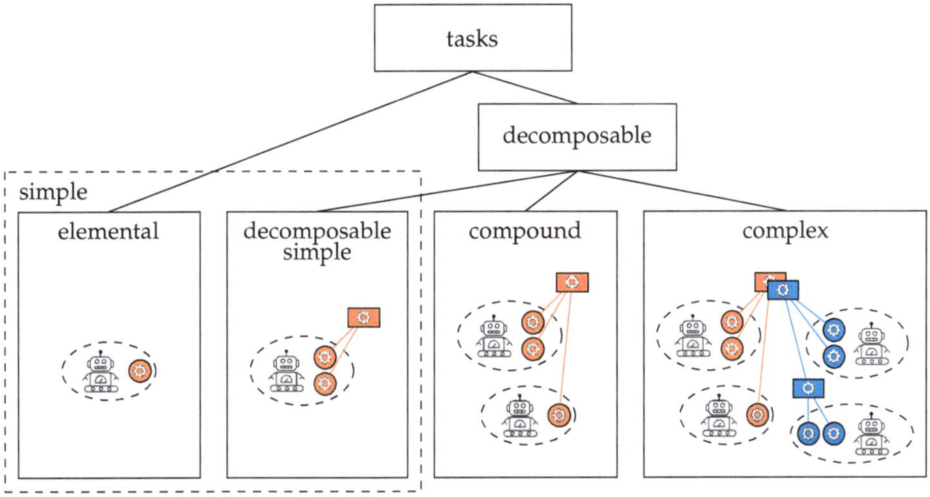

Figure 2.1: Relations between task types according to Zlot [Zlo06]. The dashed ellipses express a potential valid allocation of tasks to agents. Colored circles depict elemental tasks. Colored rectangles represent decomposable tasks with their decomposition into elemental tasks being indicated in the same color. For the complex task, the different colors represent different possible decompositions. The image is inspired by Korsah et al. [KSD13, Fig. 2].

tasks are given, a feasible subset of simple tasks to be performed must be determined. This corresponds to the problem of task decomposition.

Definition 2.5 (Task decomposition)

Given a compound or complex task n, task decomposition determines an unambiguous set of simple subtasks $\mathcal{N}_n$ such that task n is considered to be executed if all subtasks of $\mathcal{N}_n$ are executed.

The decomposition of compound tasks into simple subtasks is unambiguous. These simple subtasks may be related by some constraints, for example requiring some subtasks to be performed in a specific order. However, an equivalent representation of any compound task using the corresponding set of simple subtasks is possible. Therefore, in the following only simple and complex tasks are considered explicitely. Using the previous definitions, MRTA problems[3] are defined as follows.

[3] MRTA can be categorized as a problem of intentional coordination [Par98], i. e. an explicit and purposeful cooperation between the agents. However, coordinated behavior can also arise through emergent coordination, i. e. local interaction of individuals with each other and with the environment. Though, the potential of heterogeneous teams can better be exploited by intentional cooperation. [GM04]

Definition 2.6 (MRTA problem)
Let a set of agents $\mathcal{K}$, a set of simple and potentially also complex tasks $\mathcal{N}$, and corresponding solution constraints be given. The MRTA problem includes the problem of task allocation (see Definition 2.3) such that all tasks are allocated to agents capable of their execution. Depending on the problem features, the MRTA problem also tackles the problem of task scheduling and/or task decomposition (see Definitions 2.4 and 2.5).

Remark. *Solution constraints may for example include precedence or synchronization constraints.*

Different mathematical models can be used to represent MRTA problems. These models often include some *utility* or *cost* measure, since practical applications usually do not only seek to find just any solution to an MRTA problem, but a solution that is as good as possible w. r. t. some criterion. The higher the utility or the lower the cost of a solution, the more a solution is preferred over others.

Based on the problem features and the utility or cost measure, the following cases can be differentiated w. r. t. whether task scheduling and task decomposition are considered additionally to task allocation:

MRTA problem with task allocation only:
If the task set $\mathcal{N}$ contains only simple tasks and furthermore,

- the utility or cost measure only depends on the allocation of tasks to agents and not on the tasks' orders or
- given constraints define the task sequence for all tasks $n \in \mathcal{N}$ or
- the task set $\mathcal{N}$ contains only one simple task,

the MRTA problem is to define a task allocation.

MRTA problem with task allocation and task scheduling:
If the task set $\mathcal{N}$ contains only simple tasks and furthermore,

- the utility or cost measure depends on the tasks' orders and
- the sequence of all tasks $n \in \mathcal{N}$ is not predefined by given constraints,

the MRTA problem is to define a task allocation and a task scheduling. This implies that for all agents $k \in \mathcal{K}$ all tasks $n \in \mathcal{N}_k$ allocated to agent k must be sequenced.

MRTA problem with task allocation and task decomposition:
If the task set $\mathcal{N}$ contains at least one complex task and furthermore,

- the utility or cost measure does not depend on the tasks' orders or
- given constraints define the sequence of the resulting set of simple tasks for all possible decomposition of the complex tasks contained in the task set $\mathcal{N}$,

the MRTA problem is to define a task decomposition for all complex tasks contained in the task set $\mathcal{N}$ and to define a task allocation accounting for the chosen decomposition.

MRTA problem with task allocation, task scheduling and task decomposition: If the task set $\mathcal{N}$ contains more than one task out of which at least one is a complex task and furthermore, the given constraints do not define the sequence of the resulting set of simple tasks for all possible decomposition of the complex tasks contained in the task set $\mathcal{N}$, the MRTA problem is to define a task allocation, a task scheduling, and a task decomposition. This implies, that for each complex task a possible decomposition into simple tasks must be chosen, the resulting set of simple tasks must be allocated to the agents and for all agents $k \in \mathcal{K}$ all tasks $n \in \mathcal{N}_k$ allocated to agent k must be sequenced.

When utility or cost measures are given, optimization methods are among the well suited solution approaches. A general definition used within this thesis of optimization in the context of MRTA problems is given in Definition 2.7. It is based on [NW06, Chapter 1], but has been tailored to MRTA problems.

Definition 2.7 (Optimization)
Optimization is the minimization (or maximization) of a non-negative scalar objective function J of $\vec{X}$, where $\vec{X}$ represents a solution to an instance of an MRTA problem as given in Definition 2.6 and $\vec{X}$ fulfills given constraints. Constraints are given in the form of functions of $\vec{X}$, which define certain equalities and inequalities that the unknown solution $\vec{X}$ must satisfy.

Remark. *The objective function J, the representation $\vec{X}$ of a solution to a heterogeneous, time-extended MRTA problem instance with precedence and synchronization constraints, and the formulation of the constraints are formally introduced for this thesis in Section 3.1.1.*

Given utility or cost measures, a mathematical representation of these measures is given by the objective function J. The domain of $\vec{X}$ together with the constraints define the *search space*. A solution $\vec{X}$ is *globally optimal* if there does not exist any other solution $\vec{X}'$ within the search space such that $\vec{X}'$ has a better objective function value than $\vec{X}$ [NW06]. *Local optimality* describes a solution $\vec{X}$ to be at least as good as all other solutions within a certain *neighborhood* of $\vec{X}$ [NW06]. A neighborhood of $\vec{X}$ is a set that contains $\vec{X}$, which can be defined in various ways. If the objective function as well as all constraints are given in the form of linear functions of $\vec{X}$, the optimization model is denoted as *linear* [PAN23].

In general, optimization-based MRTA approaches can be classified into *exact* and *approximative* approaches [CGLL23]. Exact approaches achieve globally optimal solutions by considering the whole solution space. Approximative approaches make use of heuristics or metaheuristics seeking good solutions in a computationally efficient manner. Generally, heuristics are

> "popularly known as rules of thumb, educated guesses, intuitive judgments or simply common sense. In more precise terms, heuristics stand for strategies using readily accessible though loosely applicable information to control problem-solving processes in human beings and machine(s)." [Pea84] as cited in [ZBB10, Chapter 1]

In the context of MRTA problems, in this thesis, Definition 2.8 is used for heuristics.

Definition 2.8 (Heuristic)
A heuristic is an approximative solution method for optimization problems that does not guarantee to find globally optimal solutions but seeks to find in a certain sense good solutions in a computationally efficient manner.

Besides problem specific heuristics, also metaheuristics are approximative optimization approaches. There does not exist a standardized definition of the term metaheuristic [ZBB10, Chapter 4.3]. What most metaheuristics have in common, however, is that they are defined independently of a specific optimization problem and that they furthermore seek to find a balance between *intensification* and *diversification* [ZBB10, Chapter 4.3]. Intensification describes the feature of a heuristic to generate and investigate new potential solutions by exploiting at least some properties of already visited (good) solutions [ZBB10, Chapter 4.1]. Diversification means the feature of a heuristic to explore different regions of the search space [ZBB10, Chapter 4.1]. The definition of the term metaheuristic as given in Definition 2.9 is based on the definitions given in [BM17] and [OL96] as cited by Zäpfel et al. [ZBB10].

Definition 2.9 (Metaheuristic)
A metaheuristic is a general algorithmic framework that can be applied to a variety of different optimization problems with relatively few modifications to define heuristics for the specific optimization problem under consideration. A metaheuristic combines concepts for exploring the search space (diversification) and strategies for intensification in order to efficiently find good solutions.

In general, optimization methods use the information about the specific problem instance and on the general structure of the problem to solve an instance of an optimization problem. For the application of *reoptimization* methods, an *initial problem instance* $\mathcal{I}$ and a globally optimal solution $\vec{X}^*_{\mathcal{I}}$ to the initial problem instance $\mathcal{I}$ must be known. Reoptimization seeks to solve a problem instance $\mathcal{I}_{\text{mod}}$ which can be generated by applying a slight modification to the initial problem instance $\mathcal{I}$. A schematic representation of the procedures and the connection between the information available to reoptimization methods in comparison to optimization methods is depicted in Figure 2.2. The depicted optimization is assumed to be an exact method. Based on the

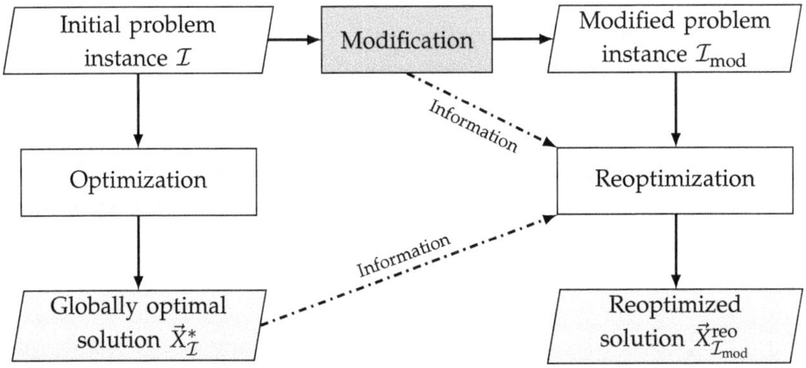

Figure 2.2: Schematic representation of the procedures and the information available to optimization and reoptimization methods.

explanations given in [BHMW08] and [ABE09], reoptimization in the context of MRTA in this thesis is generally defined as follows.

Definition 2.10 (Reoptimization)

Reoptimization solves an instance $\mathcal{I}_{mod}$ to an MRTA optimization problem (as defined in Definition 2.7) that is distinguished from a known initial problem instance $\mathcal{I}$ by a known modification applied to $\mathcal{I}$. To solve the problem instance $\mathcal{I}_{mod}$, reoptimization makes use of the knowledge of

- *a globally optimal solution $\vec{X}^*_{\mathcal{I}}$ to the initial problem instance and*
- *the modification applied to $\mathcal{I}$ to generate $\mathcal{I}_{mod}$*

in order to provide a reoptimized solution $\vec{X}^{reo}_{\mathcal{I}_{mod}}$ to the modified problem instance $\mathcal{I}_{mod}$.

Reoptimization methods are heuristic approaches and thus cannot guarantee to yield globally optimal solutions [BHMW08, ABE09].

The quality of both approximative optimization as well as of reoptimization approaches can be evaluated by the approximation ratio, if the globally optimal solution to the same problem instance is known. The following definition of the approximation ratio is based on the explanations given by Ausiello et al. [ABE09].

Definition 2.11 (Approximation ratio)
The approximation ratio α of an approximative optimization approach applied to an instance $\mathcal{I}$ of an optimization problem is defined as the ratio between the objective function value of the approximate solution $J(\vec{X}_{\mathcal{I}}^{aprx})$ and the objective function value of a globally optimal solution $J(\vec{X}_{\mathcal{I}}^{*})$ to the same problem instance, i. e.

$$\alpha = \frac{J(\vec{X}_{\mathcal{I}}^{aprx})}{J(\vec{X}_{\mathcal{I}}^{*})}. \tag{2.1}$$

For optimization problems according to Definition 2.7 that aim at minimizing (maximizing) a positive valued objective function, the approximation ratio is in $[1, \infty)$ ($[0, 1]$).

For the application of some approximative optimization approaches to specific minimization (maximization) problems, guarantees on the quality of the resulting solutions are given in the form of upper (lower) bounds on the corresponding approximation ratios. If these bounds are the smallest (largest) bounds possible, they are denoted as *tight*.

The terminology and fundamentals introduced in this section will be used in the following to present the state of the art on relevant MRTA problems, optimization and reoptimization methods. To begin with, an overview of common classifications of MRTA problems is given.

2.2 Classification of Multi-Robot Task Allocation Problems

As given in Definition 2.6, MRTA solves the problem of determining which task(s) should be executed by which agent(s) in order to achieve an overall system objective [GM04, KSD13]. Such coordination problems arise in various application domains and show multiple different characteristics. To classify MRTA problems, several taxonomies have been proposed. An overview of the relevant ones is given in this section.

Gerkey [Ger03] and Gerkey and Matarić [GM04] are the first to introduce a domain-independent taxonomy. Their taxonomy differentiates MRTA problems on three axes:

Single-task robots[4] (ST) versus multi-task robots (MT):
The first axis concerns the nature of the agents' ability to parallelize tasks. In ST problems, each agent can execute one task at a time, while in MT problems some agents are capable of executing multiple tasks simultaneously.

[4] In their taxonomy, Gerkey and Matarić [GM04] use the term *robot* for what is called *agent* (see Definition 2.2) in this thesis.

Single-robot tasks (SR) versus multi-robot tasks (MR):
The second axis differentiates MRTA problems according to the nature of the tasks. MRTA problem instances with SR consider only tasks that require exactly one agent for their execution, while MR problem instances take into consideration some tasks that need more than one agent to be executed.

Instantaneous assignment (IA) versus time-extended assignment (TA):
The third axis distinguishes MRTA instances according to the temporal availability of information of the agents, tasks and their environment. In MRTA instances with IA, the allocation of tasks to the agents is done instantaneously since no information that allows for planning future allocations is available. In contrary, in MRTA problems with TA, also a planning of future allocations is done. This requires information on the set of all tasks to be performed by the agents to be available.

To denote the nature of an MRTA problem, Gerkey and Matarić use a triplet of the introduced abbreviations. For example, ST-MR-TA describes a problem with single-robot tasks, multi-task robots and time-extended assignment. In their taxonomy, tasks are not categorized but understood as a "subgoal that is necessary for achieving the overall goal of the system, and that can be achieved independently of other subgoals (i. e. tasks)" [GM04]. In accordance with this assumption of tasks being independent of each other, they define the utility of a solution to only dependent on the sum of individual agent-task assignments and not on scheduling, i. e. the order of the tasks. Thus, both problems with interrelated utilities, i. e. utilities that depend on the overall allocation of tasks to agents, as well as problems with temporal constraints among tasks, such as constraints on sequential or parallel task execution, are not captured by the taxonomy.

To overcome these limitations, Korsah, Stentz and Dias [KSD13] present an extension of the taxonomy of Gerkey and Matarić [GM04] which they call *iTax*. It explicitly takes into consideration both interrelated utilities as well as temporal task constraints and accounts for the different types of tasks as defined by Zlot [Zlo06] that are presented in Section 2.1. The *iTax* taxonomy consists of two levels. The first level denotes the so-called "degree of interrelatedness" [KSD13] for which four categories are defined. The second level called, "problem configuration" [KSD13], further defines the nature of the problem using the taxonomy of Gerkey and Matarić [GM04]. The four degrees of interrelatedness are defined as:

No dependencies (ND):
For MRTA problems falling into the ND category, the utility of every agent-task assignment is independent of all other tasks and agents. Both simple and compound tasks can be considered as long as their utilities are not interrelated. Schedule optimization is irrelevant, since the order of task execution does not influence the overall utility.

In-schedule dependencies (ID):
Problems of the ID category consider simple or compound tasks for which the utility of assigning a task to an agent depends on the other tasks assigned to that agent, i. e. there exist "intra-schedule" dependencies. Scheduling optimization must be considered for these problems, but there is no coupling of the scheduling optimization problems of the individual agents.

Cross-schedule dependencies (XD):
The XD category comprises MRTA problems in which the utility of assigning a task to an agent depends not only on other tasks assigned to that agent but also on the schedules of other agents, i. e. there exist "inter-schedule" dependencies. These inter-schedule dependencies may e. g. be caused by constraints between schedules of different agents such as precedence or synchroniation constraints between tasks assigned to different agents. Consequently, the optimization of the schedules of individual agents are coupled problems.

Complex dependencies (CD):
If additionally to simple and compound tasks also complex tasks are to be considered, the problem falls into the CD category. The overall utility depends not only on the schedules of all agents, but also on the chosen decompositions of the complex tasks, i. e. the task decomposition problem is coupled with both task allocation and task scheduling problems.

To denote the nature of an MRTA problem, Korsah, Stentz and Dias [KSD13] use the above defined abbreviations describing the level of interrelatedness followed by the categorization of the second level given in square brackets. For example, XD [ST-SR-TA] denotes a time-extended problem with cross-schedule dependencies, single-task robots and singe-robots tasks. While ND problems can be reformulated as a linear assignment problem and are solvable in polynomial time, the problems of the categories ID, XD and CD are all NP-hard. [KSD13]

Another extension to the taxonomy of Gerkey and Matarić [GM04], that is independent of the iTax taxonomy of Korsah [KSD13], has been proposed by Landén et al. [LHD12]. They introduce four more axes on which to differentiate MRTA problems. *Unrelated utilities vs. interrelated utilities* differentiate whether the utility function depends only on the sum of individual agent-task-assignments or whether other task allocations influence the utility associated with an assignment. Whether constraints between tasks have to be considered is indicated by the axis distinguishing *independent tasks vs. constrained tasks.* If solving the MRTA problem is separated from task execution, they denote it as *external allocation view* whereas in problems with *internal allocation view,* task allocation as well as task execution are performed by an agent or several agents of the respective multi-robot system. To account for different task allocation environments, they differentiate between *static allocation environment,* where all relevant information is known in advance, and *dynamic allocation environment,* where dynamic changes concerning e. g. constraints or agents may occur. Despite covering many aspects, this taxonomy is not widely used.

Nunes et al. [NMMG17] present another extension of the taxonomy of Gerkey and Matarić [GM04]. For time-extended MRTA problems, it adds another axis to the taxonomy focusing on temporal and ordering constraints. The additional axis differentiates between different types of constraints:

Time window versus synchronization and precedence: While MRTA problems with time window constraints define a feasible time interval for task execution, synchronization and precedence constraints restrain the ordering of tasks. Precedence constraints usually require an individual task to be executed before the execution of another task may start, and synchronization constraints between a subset of tasks enforce their simultaneous start of execution.

Within these subcategories, if applicable, it is furthermore distinguished between hard temporal constraints, which must be satisfied, and soft temporal constraints, which allow for some violation at an expense of some penalty. Additionally, w. r. t. the consideration of uncertainty[5], they differentiate deterministic and stochastic models if applicable. [NMMG17]

Gini [Gin17] also focuses on temporal and ordering constraints within MRTA problems and denotes this class of MRTA problems as *MRTA/TOC*. While different temporal models are considered, including time windows as well as precedence and synchronization constraints, no taxonomy distinguishing different kinds of temporal and ordering constraints is proposed.

Due to the comprehensiveness and prevalent acceptance of the iTax taxonomy of Korsah et al. [KSD13], it is applied in this thesis. As outlined in Section 1.2, this thesis focuses on time-extended MRTA, i. e. task allocation as well as task scheduling is considered. Agents are assumed to be single-task robots. The model under consideration (see Section 3.1.1) assumes single-robot tasks. However, the introduction of synchronization constraints to bundle several tasks allows for the modeling of multi-robot tasks. The consideration of synchronization and precedence constraints between tasks results in cross-schedule dependencies. Thus, using the iTax taxonomy [KSD13], the heterogeneous, time-extended MRTA problems with precedence and synchronization constraints considered in this thesis fall into the XD [ST-MR-TA] category.

Before the presentation of the state of the art on methods to solve time-extended MRTA problems, a short introduction to the multiple traveling salesman problem (MTSP) and the vehicle routing problem (VRP) is given. These problems share relevant features with time-extended MRTA [AH17, SKRJ20]. Due to these similarities, models for time-extended MRTA problems are often based on MTSP or VRP formulations [BHK13].

---

5 Exemplary sources of uncertainty are e. g. durations for tasks' completion and agents' transitioning between tasks. However, most MRTA models are deterministic and uncertainty is dealt with at execution time [NMMG17].

2.3 Related Problems: Multiple Traveling Salesman and Vehicle Routing

Since the *multiple traveling salesman (MTSP)* and the *vehicle routing problem (VRP)* also consider the allocation and scheduling of distributed locations to a set of mobile agents, they are closely related to time-extended MRTA problems.

The MTSP is a generalization of the well-known traveling salesman problem (TSP). The TSP is concerned with, given a set of cities and the costs of travel between all cities, finding the cost optimal tour through all cities and returning to the starting point. In the MTSP, the cities can be distributed between a given number of several salesmen which all start and end their tours at a single depot. The TSP as well as the MTSP optimization problem are both NP-hard. [Dav10, CK21]

Closely related to the MTSP is the vehicle routing problem (VRP). The VRP is a generalization of the MTSP [CK21]. The VRP addresses the execution of a set of transportation requests by a fleet of vehicles. Given the set of vehicles and the set of customers associated to given transportation requests, the goal is to determine a feasible set of vehicle routes such that all (or a subset of) transportation requests are fulfilled at minimum cost [TV02, ITV14]. The VRP, which was first introduced by Dantzig and Ramser [DR59] in 1959, is a widely studied problem in operations research [ITV14]. The basic version of the VRP is the *Capacitated VRP (CVRP)*, in which a homogeneous fleet of vehicles, all starting and ending their route at one single depot, fulfill given customer delivery demands subject to capacity constraints of the vehicles. Several problem variants of the VRP considering different kinds of constraints have been proposed out of which some important ones include time windows, i. e. restricted time periods for serving each customer, the consideration of both pickup and delivery requests, or multi-depot problems with individual depots for the vehicles. [TV02, ITV14]

The MTSP has also been generalized to problems of various other application domains apart from transportation and delivery. Examples include the *home care crew scheduling problem (HCCSP)*, i. e. the problem of assigning and scheduling caretakers to patients at different locations [BF06, KLB09, RJDL12, ABLW13], or the problem of defining service routes for technicians serving different customers, the so-called *technician routing and scheduling problem (TRSP)* [KPDH12, PGM13, ZS17]. Table 2.1 gives an overview of the terminology for agents and tasks used in the problems related to MRTA.

Table 2.1: Overview on the common terminology in MRTA and related problems.

MRTA	TSP	MTSP	VRP	HCCSP	TRSP
agents	one salesman	salesmen	vehicles	caretakers	technicians
tasks	cities	cities	customers	patients	customers

Time-extended MRTA problems can be modeled as generalizations of the MTSP [AH17, CK21]. Since also VRPs are a generalization of the MTSP, VRP and MRTA problem formulations can be similar. For example, Korsah [Kor11] adapts a VRP formulation to model time-extended MRTA problems considering various constraints. An introduction to two commonly used mathematical MTSP formulations is given in Appendix A.1.

In the following section, an overview of optimization-based MRTA solution methods is given. Both approximative and exact solution methods are investigated.

2.4 Optimization-Based Solution Approaches

Optimization-based solution approaches for time-extended MRTA problem instances model them as discrete optimization problems and seek to find agent-task allocations and task schedules such that a given optimization objective is optimized and the resulting allocations and schedules respect given constraints (see Definition 2.7). In general, optimization-based MRTA approaches can be classified into exact and approximative approaches [CGLL23].

Exact approaches achieve globally optimal solutions by considering the whole solution space. Due to the strong NP-hardness of the time-extended MRTA problem [GM04], they become intractable with growing problem size [APP$^+$22]. Additionally, the consideration of constraints such as precedence and synchronization can significantly increase the computation time [KKB$^+$12]. Thus, even though they perfectly meet the requirement of high quality solutions, they are not suited for application in interactive MRTA optimization systems, since they cannot meet the requirement of responsiveness (see Section 1.1)[6]. Consequently, they are not considered in the following. However, an overview on exact approaches for time-extended MRTA and related problems with cross-schedule dependencies is given in Appendix B.1.

Approximative approaches make use of heuristics or metaheuristics seeking good solutions in a computationally efficient manner. An overview on existing approximative solution approaches for time-extended MRTA problems is given in the following. Since approximative approaches generally do not solve to optimality, the presented approaches are classified on whether they give guarantees on the solution quality whilst being capable of solving time-extended MRTA problems with the required features, i. e. time-extended MRTA problems with heterogeneous teams of agents as well as precedence and synchronization constraints.

[6] In the evaluation conducted in this thesis, MRTA problem instances without precedence and synchronization constraints with up to eight tasks and three agents could be solved within 10 s by the applied exact solution approach. For problem instances of the same size considering one or more precedence or synchronization constraint, the exact solution approach required considerably more calculation time (see Chapter 5).

Both heuristic (see Definition 2.8) and metaheuristic (see Definition 2.9) approaches have been introduced to approximately solve time-extended MRTA optimization problems. Different *constructive heuristics*[7] and *improvement heuristics*[8] have been applied. For example, Mitiche et al. [MBG15] present three different greedy constructive heuristics and an improvement heuristic which account for heterogeneous capabilities of agents and task deadlines. However, no guarantee on the resulting solution quality is given.

Metaheuristics can be categorized in trajectory-based and population-based methods. Trajectory-based metaheuristics use a single solution that is modified throughout the search process. Improvements are always accepted while impairments might be accepted with a certain probability in order to escape local optima. Population-based metaheuristics use populations, i. e. multiple candidate solutions at a time, during the search process. [KHE15, APP+22]

Examples for trajectory-based metaheuristics include for example tabu search[9] (TS), simulated annealing[10] (SA) and large neighborhood search[11] (LNS) [EA20]. Mosteo and Motano [MM06] apply SA to a time-extended MRTA problem for heterogeneous agents that includes task decomposition. SA is also used by David and Rögnvaldsson [DR21] as well as Wang and Chen [WC22] to coordinate a team of agents. Also TS has been applied to solve time-extended MRTA problems [AKH03, ZXS14]. Booth et al. [BNB16] use LNS to improve initially generated MRTA solution for social robots applied in a retirement home.

7 Constructive heuristics build a solution step by step using defined rules [Sal17, Chapter 2].

8 Improvement heuristics try to improve a solution via a selected neighborhood [Sal17, Chapter 2].

9 The tabu search algorithm uses a short-term memory called tabu list. In every iteration, using a defined neighborhood operator, all neighboring solutions to the current one are generated. The best solution within this neighborhood, which is not found in the tabu list, is chosen to be the next solution and the current solution is added to the tabu list. A move to a new solution is always executed even if the current solution is locally optimal. This allows to escape local optima and the tabu list prevents short term cycling. [ZBB10, Chapter 6.2]

10 In simulated annealing, a candidate solution out of the neighborhood of the current solution is chosen randomly. If it improves the current solution, it is accepted. But even if it has a worse objective value than the current solution, it is accepted with a probability greater 0 which increases the smaller the objective value difference to the current solution is. Over time, a parameter t, called temperature, is decreased, which reduces the probability of accepting worse solutions. The probability for solutions whose objective function value is worse than the one of the current solution and differs from it by ΔJ, is defined as $p(\Delta J) = e^{-(\Delta J/t)}$. [ZBB10, Chapter 6.4]

11 In large neighborhood search, a destroy and a repair method implicitly defines a solution neighborhood. A part of the current solution is destructed by the destroy method and subsequently rebuild by a repair method. The destroy method typically contains stochastic elements such that different parts of the solution are destroyed with every call of the method. [PR19]

Population-based metaheuristics, including genetic algorithm[12] (GA), memetic algorithm (MA)[13], ant colony optimization[14] (ACO), and particle swarm optimization[15] (PSO), have also been applied to time-extended MRTA problems. Especially GAs are wide-spread and applied in various application domains including for example disaster response [JDS11] and thermosolar plant monitoring [MFGC21]. Padmanabhan Panchu et al. [PRSB18] propose a GA to solve time-extended MRTA with cooperative tasks and precedence constraints. A MA that considers tasks that need either one or two agents to be executed is given by Liu and Kroll [LK15]. Yakici [Yak16] uses ACO to coordinate a fleet of unmanned aerial vehicles (UAVs) with limited flight time to visit as many target positions as possible. For solving a homogeneous time-extended MRTA problem considering two objectives, Chen et al. [CZDL18] combine a bi-objective ACO algorithm with a sequential variable neighborhood search (VNS)[16]. PSO is combined with evolutionary game theory[17] (EGT) by Zhu et al. [ZTY17] to solve homogeneous time-extended MRTA problems. Wei et al. [WJC20] propose a PSO approach that considers two optimization objectives in order to balance the workload between the agents of the robotic team which has to retrieve a set of distributed objects.

Also in the area of vehicle routing, heuristics and especially metaheuristics are widely prominent solution approaches [EAR23]. A comprehensive review thereon is given by Elshaer and Hadeer in [EA20]. They classify 299 publications proposing metaheuristic approaches for VRPs and analyze the frequency of the usage of the different metaheuristics.

Despite the large number of heuristic and metaheuristic approaches in literature, only few approximative solution methods with performance guarantees exist. Performance guarantees on the solution quality for approximative approaches to optimization problems seeking to minimize an objective function, are usually given in terms of upper

[12] Genetic algorithms were inspired by the process of evolution. In each iteration, a selection operator is used to choose some solutions out of the current population. A crossover operator is applied to generate new solutions by merging properties of the selected solutions. To some percentage of the new solutions, a random perturbation is applied, called mutation. After evaluating the newly generated solutions, a new population is generated using a replacement operator which defines, which solutions of the newly generated ones and the old population form the new population. [ZBB10, Chapter 7.1]

[13] Memetic algorithms combine evolutionary algorithms like GA with local search procedures. [NC12]

[14] The idea of ant colony optimization is derived from the behavior of ant colonies. Artificial ants construct solutions by putting together several solution components. For each solution component, transitions to other components are defined which determine the candidates for the next construction step. For every solution component chosen by an ant, it leaves a pheromone trace which vanishes over time. The probability for any ant to transition to a solution component is proportional to the pheromone trail of the component. [ZBB10, Chapter 5.2]

[15] In particle swarm optimization, various particles, i. e. solutions, are distributed in the search space. Each solution evaluates its value of the objective function and furthermore has information about the own best solution found so far, the best solution of the whole swarm and its own velocity which depends on the two previous values. In every iteration, each particle moves with its own velocity towards the best solution of the whole swarm and updates all information. [ZBB10, Chapter 8.2.4]

[16] In variable neighborhood search, several neighborhood operators are alternatingly used. This procedure aims at overcoming local optima. [GP10, Chapter 3]

[17] Evolutionary game theory combines the mathematical models of game theory with the basic concept of Darwinism to model time evolution of populations. [Tan15]

bounds on the resulting approximation ratio (see Definition 2.11). Due to the very limited number of approximative solution approaches giving performance guarantees in the MRTA problem domain, also related problem domains are considered in the following.

Yadlapalli et al. [YJRD11] consider a multiple depot hamiltonian path problem (MD-HPP) with symmetric costs that satisfy the triangle inequality for unmanned vehicles (UVs). In the considered MD-HPP a set of agents must visit a given set of locations such that the overall costs are minimized. The locations can either be visited by any agent or only by a specific one. All agents start at individual depots and do not have to return to their depot at the end of their tour. The authors prove their approximative algorithm to yield an approximation ratio bounded above by $^{11}/_{3}$. For every UV, additionally to an initial start depot also a terminal position is given. They furthermore consider the UVs to be heterogeneous in such a way that targets can either be visited by all UVs or only by a specific one.

Also for the multiple depot MTSP (MD-MTSP) with symmetric edge costs that satisfy the triangle inequality some heuristics with performance guarantees have been proposed. MD-MTSP refers to a variant of the MTSP in which each agent starts and ends its tour at an individual depot. Malik et al. [MRD07] consider the generalized MD-MTSP. In the generalized problem considered by the authors, only a defined maximum number of the agents can be chosen to visit the destinations. The algorithm yields an approximation ratio bounded above by 2. Xu et al. [XXR11] propose an extension to the Christofides algorithm[18] to solve the MD-MTSP and prove a tight approximation ratio of $2 - {}^{1}/_{k}$, k being the number of agents. Yadlapalli et al. [YRD10] investigate a special case of the MD-MTSP considering two agents that can differ either in travel velocity or in their capability of visiting destinations. In case of different capabilities, destinations are split into three disjoint subsets: one subset of destinations for each agent and one common subset of destinations which can be visited by either of the two agents. They prove their proposed heuristic to yield approximation ratios bounded above by 3.

In the application domain of time-extended MRTA problems Zhang and Parker [ZP13c] introduce two simple greedy constructive heuristics as well as two more complex ones that work with reformulations of the problems as assignment problems. They consider MRTA problems with tasks that require direct cooperation between agents. Upper bounds on the approximation ratios for three of the four approximation heuristics are given. However, the upper bounds either grow linearly with the number of tasks considered or depend on potential conflicts for agents between different agent cooperations[19] in the specific problem instance considered. If the objective function of MRTA problems is defined such that it does not depend on task scheduling but only on the allocation of tasks to agents, performance guarantees have also been given

18 The Christofides algorithm is an approximate approach for the TSP with metric distances. It is based on the computation of the minimum spanning tree of the complete graph representing the TSP and guarantees an upper bound on the resulting approximation ratio of $\alpha \leq {}^{3}/_{2}$. [Chr22]

19 These conflicts arise since an agent can only be part of one cooperation at a time.

[LCS11, LCS12, LCS13, LCS15a], but these publications do not consider time-extended MRTA problems.

An overview of the presented approximative optimization approaches is given in Table 2.2, which depicts the solution approach, problem domain, problem features and performance guarantees given for the publication. Most heuristic and metaheuristic solution approaches for time-extended MTRA problems lack guarantees on the quality of the approximate solutions generated. This holds especially if heterogeneous agents or temporal constraints are considered [Gin17]. For the majority of metaheuristic approaches, constant guarantees are difficult to obtain, since most metaheuristics are non-deterministic [DRAR15]. Constant guaranteed approximation ratios are mostly given for heuristics solving problems without temporal constraints that either consider homogeneous agents [MRD07], a limited number of agents [YRD10] or with tasks that can be either only executed by one or by all agents [YJRD11]. In summary, no performance guarantee can be given for an approximative optimization approach solving heterogeneous, time-extended MRTA problems with precedence and synchronization constraints.

Furthermore, optimization-based solution approaches typically consider each problem instance individually. Thus, solution stability, another requirement on solution approaches for the desired interactive MRTA optimization system (see Section 1.1), is not considered by optimization approaches if applied to slightly modified problem instances. Contrary to this, reoptimization approaches use the knowledge of a previously generated solution when solving a slightly modified problem instance (see Definition 2.10), which is a promising approach w. r. t. solution stability. An overview of relevant existing reoptimization methods is given in the following section.

Table 2.2: Overview of approximative solution approaches to solve time-extended MRTA or related optimization problems. The depicted approaches all consider a time-extended planning horizon. For each publication, the applied solution approach and the problem domain are given. Furthermore, it is indicated, whether heterogeneous teams of agents (Het.), precedence constraints (Prec.) and synchronization constraints (Sync.) are considered. Whether an upper bound on the resulting approximation ratio α is given is indicated in the last column.

Publication	Solution approach	Problem domain	Het.	Prec.	Sync.	Upper bound on α
[MBG15]	constructive & improvement heuristic	MRTA	✓	✗	✗	not given
[MM06]	SA	MRTA	✓	✗	✗	not given
[DR21]	SA	MRTA	✗	✗	✗	not given
[WC22]	SA	MRTA	✓	✗	(✓)	not given
[AKH03]	TS	MRTA	✗	✓	✗	not given
[ZXS14]	TS	MRTA	✗	✗	✗	not given
[BNB16]	LNS	MRTA	✓	✓	✓	not given
[JDS11]	GA	MRTA	✓	(✓)	✗	not given
[LK12]	GA	MRTA	✗	✗	✗	not given
[PRSB18]	GA	MRTA	✗	✓	(✓)	not given
[MFGC21]	GA	MRTA	✓	✗	✗	not given
[LK15]	MA	MRTA	✓	✗	✓	not given
[Yak16]	ACO	MRTA	✗	✗	✗	not given
[CZDL18]	ACO with VNS	MRTA	✗	✗	✗	not given
[ZTY17]	PSO with EGT	MRTA	✗	✗	✗	not given
[WJC20]	PSO	MRTA	✗	✗	✗	not given
[YJRD11]	heuristic algorithm	MD-HPP	(✓)	✗	✗	$11/3$
[MRD07]	heuristic algorithm	generalized MD-MTSP	✗	✗	✗	2
[XXR11]	extended Christofides algorithm	MD-MTSP	✗	✗	✗	$2 - 1/k$ (k: number of agents)
[YRD10]	heuristic algorithm	MD-MTSP (two agents)	(✓)	✗	✗	3
[ZP13c]	constructive heuristics	MRTA	✓	✗	(✓)	grows linearly with problem features

2.5 Reoptimization Approaches

Since reoptimization approaches seem promising w. r. t. the requirement of solution stability (see Section 2.4), an overview of existing reoptimization methods is given in this section. To evaluate whether also the requirement of guaranteed high quality solutions (see Section 1.1) can be met, the presented approaches are analyzed w. r. t. performance guarantees.

Approaches dealing with problem modifications in the domain of MRTA problems are investigated in Section 2.5.1. This investigation reveals that no centralized reoptimization approaches for time-extended MRTA problems with performance guarantees, as required for the intended interactive application (see Section 1.2), exist. Conrary to this, reoptimization heuristics with guarantees on the solution quality have already been proposed for the TSP, which corresponds to the special time-extended MRTA of one agent having to perform all tasks. The respective approaches and their corresponding guarantees are presented in Section 2.5.2.

2.5.1 Reoptimization in Time-Extended MRTA and Related Problems

Reoptimization for time-extended MRTA is a rather new field of research and to the best of the author's knowledge no centralized reoptimization approaches for time-extended MRTA problems exist. However, some decentralized approaches considering modifications to MRTA problem instances have been proposed.

For example, market-based approaches have been applied to incorporate newly arriving tasks during mission execution, e. g. [STBE09, HP13, NG15, Lee18, LLW$^+$18]. Since agent-task-allocations are evaluated locally by each individual agent in market-based approaches, these approaches make no assumption on the heterogeneity or homogeneity of agents [Lee18]. However, these approaches do not give any performance guarantees. Furthermore, it has been shown for some auction-based methods that their solution quality deteriorates if dynamic allocations over time are considered compared to static problems where all tasks are known in advance [SSPÖ15, GC22].

A decentralized approach to react to newly arriving tasks during mission execution is proposed by Buckman et al. [BCH19]. They propose an extension to the distibuted consensus-based bundle algorithm (CBBA) [CBH09] which they call CBBA with partial replanning (CBBA-PR). In CBBA, a bundle building phase in which agents iteratively generate their schedules by bidding on tasks, is followed by a consensus phase in which agents resolve misunderstandings regarding the allocations of each task [CBH09]. Upon arrival of a new task in CBBA-PR, each agent releases a part of its allocated tasks for reallocation. Thus, parts of the initially generated solutions are kept unchanged. This approach however, does not allow for performance guarantees w. r. t. the quality of the solution generated.

A slightly different problem is considered by Emam et al. [EMN+20]. They consider the case that agents' capabilities during mission execution differ from the anticipated ones. By constantly comparing expected execution costs against observed ones during execution, specialization parameters representing agents' capabilities are constantly updated during task execution. Since they consider an instantaneous assignment problem, no task scheduling is considered.

Another method that incorporates new tasks is presented by Ghassemi and Chowdhury [GC22]. They consider an MRTA problem with task deadlines as well as agent range and payload constraints to which new tasks are added during mission execution. Their decentralized approach is based on constructing and matching bipartite graphs. However, each agent only selects the next task to execute at a time, and therefore no task scheduling is taken into consideration.

Also in the application domain of vehicle routing, so-called dynamic VRPs (DVRP) deal with reacting to dynamically emerging or changing information on the problem instance [OXM+21]. Many approaches to solve such problems have been proposed, out of which only some use reoptimization methods [OXM+21]. Comprehensive reviews on DVRPs are given by Pillac et al. [PGGM13] and by Ojeda Rios et al. [OXM+21]. However, to the best of the author's knowledge, no relevant reoptimization-based DVRP approaches with guarantees on the solution quality exist.

Table 2.3 summarizes the publications presented in this section. For each publication, the considered modification, the mode of solution computation (centralized vs. distributed), the considered planning horizon (IA vs. TA) as well as whether performance guarantees are given, are depicted. Overall, no reoptimization approach for time-extended MRTA problems exists that centrally evaluates all information on the modified problem instance. Existing approaches for incorporating problem modifications either build upon distributed local information ([STBE09, HP13, NG15, Lee18, LLW+18, BCH19, GC22]) or only consider instantaneous assignment problems, i. e. do not incorporate task scheduling ([STBE09, LLW+18, EMN+20, GC22]). Furthermore, no guarantees on the resulting solution quality is given for any of these approaches.

In contrast to this, for the TSP, i. e. the problem of optimally routing one agent to visit several tasks, centralized reoptimization approaches with guarantees on the resulting solution quality have been proposed. These are discussed in the following section.

2.5.2 Reoptimization of the Traveling Salesman Problem

If only one agent to perform all given tasks is considered in a time-extended MRTA, this special case corresponds to the TSP. For the TSP, several reoptimization heuristics

Table 2.3: Overview of dynamic MRTA solution approaches accounting for modifications in the problem instance. For each publication, the applied modification is given. It is furthermore indicated, whether the approach is based on locally distributed or centralized information and whether time-extended (TA) or instantaneous assignments (IA) are considered. Moreover, it is indicated, whether heterogeneous teams of agents (Het.), precedence constraints (Prec.) and synchronization constraints (Sync.) are considered. Whether an upper bound on the resulting approximation ratio α is given is indicated in the last column.

Publication	Modification	Solution computation	Planning horizon	Het.	Prec.	Sync.	Upper bound on α
[STBE09]	environmental changes, task insertion, task deletion	distributed	IA	✗	✗	✗	not given
[HP13]	task insertion	distributed	TA	(✓)	✗	✗	not given
[NG15]	task insertion	distributed	TA	(✓)	✗	✗	not given
[Lee18]	task insertion	distributed	TA	(✓)	✗	✗	not given
[LLW+18]	agent failure	distributed	IA	(✓)	✗	✗	not given
[BCH19]	task insertion	distributed	TA	(✓)	✗	✗	not given
[EMN+20]	capability variation	centralized	IA	✓	✗	✗	not given
[GC22]	task insertion	distributed	IA	✓	✗	✗	not given

with performance guarantees in the form of upper bounds on the resulting approximation ratio exist.[20]

Archetti et al. [ABS03] consider the reoptimization problem of adding a new node, i. e. a new task, to a previously exactly solved TSP instance. They prove the corresponding reoptimization problem to remain NP-hard and propose the so-called *cheapest insertion procedure* to solve the reoptimization problem. The resulting approximation ratio is proven to be bounded above by $3/2$ in the case of metric distances. This bound is shown to be tight, i. e. no smaller bound could possibly be given for the proposed heuristic.

The same problem is inspected by Ausiello et al. [AEMP09] who propose a combination of the cheapest insertion procedure and Christofides algorithm [Chr22]. They prove the resulting approximation ratio to be bounded above by $4/3$. For the insertion of $l > 1$, $l \in \mathbb{N}$, nodes, they apply a similar algorithm that chooses the best solution between Christofides algorithm and iterative cheapest insertion of the new nodes in ascending order according to their proximity to existing nodes. The proposed heuristic has an approximation ratio bounded above by $3/2 - 1/(4l+2)$. [AEMP09]

[20] In addition to the presented TSP reoptimization approaches, also reoptimization in the context of TSPs that aim at maximizing the sum of edge weights of the chosen solution [AEMP09, BH09] and for TSPs with deadlines [BKK09] have been investigated. Since a minimization problem without deadlines is considered in this thesis, these approaches are not discussed in detail.

The result of Ausiello et al. [AEMP09] on the insertion of a single node is improved by a reoptimization heuristic proposed by Monnot [Mon15] that also yields an upper bound of 4/3 on the approximation ratio, but at a lower computational complexity.

The asymmetric version of the reoptimization problem of inserting a task to a TSP is studied by Ausiello et al. in [ABE09]. In the asymmetric TSP, the distances are only required to obey the triangle inequality, but do not have to be symmetric. They prove a tight upper bound of 2 on the approximation ratio for a very simple insertion heuristic.

Also the problem of deleting a node, i. e. a task, from a TSP instance has already been studied. Archetti et al. [ABS03] prove the corresponding reoptimization problem to be NP-hard and propose a simple task deletion heuristic that skips the deleted tasks and keeps the order of the remaining ones. They can prove a tight upper bound of 3/2 on the approximation ratio for metric TSPs.

The corresponding reoptimization problem of deleting a task from an asymmetric TSP for the same deletion heuristic is studied by Aussiello et al. [ABE09] who prove the respective heuristic to yield solutions that cannot have approximation ratios greater than 2.

Böckenhauer et al. [BFH+06, BHMW08] consider the variation of a single edge cost in a TSP[21] and prove the respective reoptimization problem to be NP-hard. For the case of metric edge costs, they provide an algorithm having an approximation ratio bounded above by 7/5. Berg and Hempel [BH09] improve the result given by Böckenhauer et al. [BFH+06, BHMW08] using an algorithm that chooses the better solution between the initial one and the approximative solution generated by Christofides algorithm [Chr22] and prove the approximation ratio to have an upper bound of 4/3.

A summary on the presented TSP reoptimization results is depicted in Table 2.4. Overall, constant approximation ratios can be guaranteed for the modifications of task insertion, task deletion and edge cost variation out of which some approximation ratios are furthermore proven to be tight. Additionally, these approaches all have a polynomial complexity. Consequently, they fulfill the requirements of computational efficiency and guaranteed solution quality which are required for the intended interactive setting (see Section 1.1). Furthermore, the modified solutions generated based on the initial solution often differ from the initial solution in a deterministic manner such that the initial and the modified solutions share many features. However, this cannot be guaranteed for the heuristics that include the Christofides algorithm [Chr22] since here solutions to the modified problem might be generated from scratch. Overall, deterministic reoptimization heuristics are promising w. r. t. the requirements for the intended interactive setting (see Section 1.1). However, so far only TSPs, which correspond to time-extended MRTAs with one agent performing all tasks without any temporal constraints, have been considered.

---

21 In the context of the TSP this means that the time required by the salesman to travel between two specific cities changes, e. g. due to altered traffic conditions.

In summary, there exist promising reoptimization results for TSPs that give guarantees on the resulting approximation ratios for the problem modifications of task insertion, task deletion and edge cost variation. However, for time-extended MRTA problems so far no centralized reoptimization approaches with performance guarantees exist. In order to define solution approaches that meet the requirements of high quality solutions, responsiveness and solution stability, as defined in Chapter 1.2, the research questions to be answered in this thesis are presented in the following section.

Table 2.4: Reoptimization approaches with upper bounds on the resulting approximation ratios for the traveling salesman problem. By definition of the TSP, these approaches consider time-extended assignment for one agent and neither precedence nor synchronization constraints. For each publication, the applied modification is given, and it is indicated whether metric or asymmetric distances are considered. Furthermore, the upper bound on the approximation ratio α is given, and the last column indicates whether the upper bound is tight.

Publication	Modification	Problem characteristics	Upper bound on α	Tight bound
[ABS03]	task insertion	metric TSP	$\alpha \leq 3/2$	✓
[AEMP09]	task insertion	metric TSP	$\alpha \leq 4/3$	not analyzed
[Mon15]	task insertion	metric TSP	$\alpha \leq 4/3$	not analyzed
[ABE09]	task insertion	asymmetric TSP	$\alpha \leq 2$	✓
[ABS03]	task deletion	metric TSP	$\alpha \leq 3/2$	✓
[ABE09]	task deletion	asymmetric TSP	$\alpha \leq 2$	not analyzed
[BFH+06]	edge cost variation	metric TSP	$\alpha \leq 7/5$	not analyzed
[BH09]	edge cost variation	metric TSP	$\alpha \leq 4/3$	not analyzed

2.6 Research Questions

On the basis of the literature review given in the previous sections, in this section, the research questions to be answered in this thesis and the corresponding research contributions are presented.

As revealed by the literature overview, while there exist promising reoptimization heuristics for different modifications of the TSP, so far no reoptimization heuristics for time-extended MRTA problems have been proposed. The first research question is concerned with closing this gap.

Research Question 1:
Can centralized reoptimization heuristics be defined for modifications of the heterogeneous, time-extended MRTA problem with precedence and synchronization constraints that guarantee to find a feasible solution to the modified problem instance? Is it furthermore possible for some heuristics to guarantee constant upper bounds on the approximation ratios of the solutions they generate?

Research Contribution 1:
To answer this question, several reoptimization heuristics suitable for reoptimization problems corresponding to ten fundamental modifications of heterogeneous, time-extended MRTA problems with precedence and synchronization constraints are introduced and theoretically analyzed within this thesis. The modifications under consideration comprise important modifications such as task insertion and task deletion that have already been investigated for the TSP (see Table 2.4 in Section 2.5.2) and are extended by modification possibilities that arise from the augmented problem statement (in comparison to the TSP) of heterogeneous, time-extended MRTA problems with precedence and synchonization constraints. Therefore, the investigated modifications include

- inserting a task to an MRTA instance,
- removing a task from an MRTA instance,
- changing the position of a task,
- changing the duration of a task,
- changing the capability of an agent to perform some tasks,
- changing the transition velocity of an agent,
- adding a precedence constraint between two tasks,
- adding a synchronization constraint between two tasks,
- removing a precedence constraint between two tasks, and
- removing a synchronization constraint between two tasks.

All reoptimization heuristics proposed for the MRTA reoptimization problems belonging to these modifications ensure the resulting solution to the modified problem instance to be feasible, i. e. to respect all constraints and thus to be executable in practice. Except for the heuristics applied to the modifications of adding a precedence or synchronization constraint, all heuristics are furthermore guaranteed to find a feasible solution if the solution set of the corresponding modified problem instance is non-empty. Moreover, for the reoptimization heuristics for inserting and removing a task from a heterogeneous, time-extended MRTA problem instance with precedence and synchronization constraints, performance guarantees are given in the form of constant upper bounds on the resulting approximation ratio. Depending on the features of the time-extended MRTA instance, different bounds apply, ranging between $3/2$ and 2.

In addition to the reoptimization heuristics, also metaheuristic reoptimization approaches are of interest. Most metaheuristic solution approaches balance intensification and diversification [ZBB10, Chapter 4.3] and thus search larger areas of the search space than classical heuristics. Furthermore, as outlined in the literature review, metaheuristic approaches are widely used to solve time-extended MRTA problems. Consequently, this thesis investigates whether a combination of reoptimization heuristics with a metaheuristic solution approach is possible and whether this reveals further improvement potential.

Research Question 2:
Is a combination of the reoptimization heuristics with a metaheuristic solution approach possible for heterogeneous, time-extended MRTA problems with precedence and synchronization constraints? Is such a combined approach methodically useful and does it have the potential to further improve the solution quality?

Research Contribution 2:
The second contribution of this thesis is the introduction of a reoptimization framework that combines the proposed reoptimization heuristics with metaheuristic optimization approaches. The additional application of a metaheuristic solution approach allows for a broader search of the solution space. Any metaheuristic suitable for heterogeneous, time-extended MRTA problems with precedence and synchronization constraints can be applied. The solution generated by the respective reoptimization heuristic is used for the initialization of the metaheuristic, since the initialization of metaheuristic algorithms significantly influences their performance [LLY20, SVJ23]. It is proven, that if the metaheuristic has certain properties, the guarantees given on the feasibility and on the approximation ratios of the respective reoptimization heuristic remain valid. Furthermore, a specific realization of the proposed reoptimization framework using a genetic algorithm (GA) is introduced. The GA is specified such that it fulfills aforementioned properties and thus preserves all performance guarantees of the applied reoptimization heuristics.

Besides the theoretical results, the performance of the proposed approaches on different MRTA problem instances is of interest, which leads to the third research question to be answered in this thesis:

Research Question 3:
How do the proposed reoptimization approaches perform on heterogeneous, time-extended MRTA problem instances with precedence and synchronization constraints, and how do different problem features influence their performance? What are the advantages and disadvantages of the reoptimization heuristics compared to the GA-based metaheuristic reoptimization approach?

Research Contribution 3:
The third contribution of this thesis is an extensive evaluation to analyze the performance of the proposed reoptimization methods compared to an exact optimization approach and a conventional GA. All requirements arising from the intended interactive application, i. e. quality of the solutions, responsiveness of the solution approaches and solution stability are analyzed via appropriate performance measures. The influence of various problem features of heterogeneous, time-extended MRTA problems with precedence and synchronization constraints on the performance measures is inspected. These features comprise the problem size, the extent of the agents' and tasks' heterogeneity, and the number of precedence and synchronization constraints considered within a problem instance. The evaluation yields evidence of the suitability of the proposed reoptimization heuristics for their application in an interactive MRTA optimization system, since the reoptimization heuristics provide promising results w. r. t. the combination of the criteria relevant for application in interactive MRTA optimization systems. Overall, both reoptimization approaches clearly outperform the investigated optimization approaches w. r. t. the considered criteria.

Since reoptimization approaches rely on globally optimal initial solutions, an exact optimization approach is necessary. Within this thesis, a branch-and-price (BnP) approach introduced by Korsah [Kor11] is applied to generate globally optimal solutions. However, to the best of the author's knowledge, so far no exact solution approach takes the problem of task decomposition into consideration. Therefore, another contribution of this thesis is the introduction of three approaches to globally optimal solve respective time-extended MRTA problems with complex dependencies, i. e. problems that consider the correlated problems of task allocation, task scheduling and task decomposition. Since the focus of this thesis lies on the MRTA reoptimization problem for the intended interactive setting without complex tasks, the corresponding contribution can be found in Appendix B.

In the following chapter, heuristic solution approaches for different problem modifications are introduced and analyzed in order to answer the first research question.

3 Heuristic Reoptimization of Time-Extended MRTA Problems

To answer the first research question, reoptimization heuristics for ten fundamental problem modifications (see listing in Research Contribution 1, Section 2.6) to time-extended MRTA problems with heterogeneous teams of agents and precedence and synchronization constraints are introduced and analyzed in this chapter.

The MRTA task insertion reoptimization problem is defined and discussed in Section 3.2. Furthermore, a sufficient criterion for the approximation ratio (see Definition 2.11) resulting from the application of a reoptimization approach to be bounded above by 2 is given. The *cheapest maximum insertion heuristic* is introduced which is proven to always fulfill the respective criterion, thus resulting in a guaranteed worst-case approximation ratio of 2, and to yield feasible solutions. Furthermore, smaller upper bounds on the approximation ratio are derived for temporally unconstrained problem configurations.

In Section 3.3, the MRTA task deletion reoptimization problem is introduced and investigated. Similarly to the task insertion reoptimization problem, it is proven that for all reoptimization approaches that fulfill two assumptions, the resulting approximation ratio is bounded above by 2. The *task deletion heuristic* is introduced which is proven to always generate feasible solutions. For a temporally unconstrained problem configuration with homogeneous agents and a single depot, the respective heuristic is shown to yield approximation ratios bounded above by $3/2$.

A definition and analysis of the MRTA reoptimization problem of task position variation is given in Section 3.4. Two solution heuristics, called *initial solution approach* and *delete-insert heuristic* are introduced and proven to always yield feasible solutions to the task position variation problem. The feasibility of these two heuristics is shown to also hold for the MRTA reoptimization problem of task duration variation, which is discussed in Section 3.5. The MRTA reoptimization problems emerging from varying the capabilities of an agent to process the tasks and from varying the velocity of an agent are introduced and discussed in Sections 3.6 and 3.7, respectively. For both modifications, the initial solution approach is proven to always yield feasible solutions.

Introducing an additional precedence or synchronization constraint to an MRTA problem instance results in MRTA reoptimization problems of precedence or synchronization constraint insertion, respectively, which are introduced and analyzed in Section 3.8. For solving these kinds of reoptimization problems, the *extended delete-insert heuristic* is introduced, which is proven to yield feasible results for both kinds of reoptimization problems. In Section 3.9, the MRTA reoptimization problems corresponding to the modifications of deleting a precedence or synchronization constraint from an MRTA problem instance are introduced, and it is proven that the initial solution approach always yields feasible solutions for both. Parts of the results on the reoptimization heuristics have been established in the master theses of Kohn [Koh21] and Hahn [Hah22], both supervised by the author.

The relevant fundamentals for the aforementioned analyses are given in the following Section 3.1.

3.1 Problem Definition and Solution Feasibility

In this section, the fundamentals relevant for the analyses of heuristic solution approaches to different MRTA reoptimization problems are given. A mathematical definition of the heterogeneous, time-extended MRTA optimization problem with precedence and synchronization constraints as well as the notification used throughout this thesis are given in Section 3.1.1. A criterion to validate the feasibility of solutions is presented in Section 3.1.2. In Section 3.1.3, a generalized reoptimization problem is given, which corresponds to the afore introduced heterogeneous, time-extended MRTA optimization problem with precedence and synchronization constraints.

3.1.1 Heterogeneous, Time-Extended MRTA Optimization Problem with Precedence and Synchronization Constraints

In this section, a mathematical definition of the heterogeneous, time-extended MRTA optimization problem with precedence and synchronization constraints, which is based on a 3-index vehicle flow formulation (see Appendix A.1), is given. An analogical formulation has been used in a journal publication by the author [BKH$^+$24].

In a heterogeneous, time-extended MRTA problem instance with precedence and synchronization constraints $\mathcal{I} = \{\mathcal{N}, \mathcal{K}, \mathcal{O}, \mathcal{V}, \mathcal{D}, \mathcal{T}, \mathcal{A}, \mathcal{P}, \mathcal{S}\}$, a set of simple tasks $\mathcal{N} = \{n_1 \dots n_N\}$, $N \in \mathbb{N}$, has to be both scheduled as well as assigned to a set of agents $\mathcal{K} = \{k_1 \dots k_K\}$, $K \in \mathbb{N}$. Each agent $m \in \mathcal{K}$ starts and ends its route at its individual depot $o_m \in \mathcal{O} = \{o_1 \dots o_K\}$ and moves with a constant velocity $v_m \in \mathbb{R}_{>0}$, i.e. $\mathcal{V} = \{v_1 \dots v_K\}$ defines the set of agents' velocities. The problem can be depicted using a complete graph $G = (V, E)$. The set of vertices $V = \mathcal{N} \cup \mathcal{O}$ contains all tasks and all depots, i.e. $|V| = N + K$. The edges E are associated with metric edge costs described by distances $\delta : V \times V \rightarrow \mathbb{R}_{\geq 0}$, i.e. $\mathcal{D} = \{\delta(i,j) \,|i,j \in V\}$ describes the set of all distances. Metric distances are symmetric, positive and fulfill the triangle equality [Jun13, p. 68]. For each task, a basic task duration is defined by the set of task durations $\mathcal{T} = \{\tau_1 \dots \tau_N\}$, $\tau_i \in \mathbb{R}_{\geq 0}$ $\forall \tau_i \in \mathcal{T}$. Together with a capability $a_i^m \in \mathbb{R}_{\geq 0}$ that is defined for each agent-task pair, i.e. $\mathcal{A} = \{a_i^m \,|i \in \mathcal{N}, m \in \mathcal{K}\}$, an agent dependent task duration d_i^m results. It is given by $d_i^m = \tau_i/a_i^m$ if $a_i^m > 0$ and otherwise set to $d_i^m = \infty$. Analogously, an agent dependent edge cost $d_{i,j}^m = \delta(i,j)/v_m$, $i, j \in V$, $m \in \mathcal{K}$, follows from the distances and the velocities defined by the problem instance

$\mathcal{I}$. It resembles the time needed by agent $m \in \mathcal{K}$ to traverse edge (i, j) associated with the distance $\delta(i, j)$. Thus, the heterogeneity of the agents is described by the set of velocities $\mathcal{V}$ and by the set of agent-dependent task capabilities $\mathcal{A}$.

The set of precedence constraints $\mathcal{P} = \{p_{i,j} \,|\, i, j \in \mathcal{N}, i \neq j\}$ contains all constraints $p_{i,j} \in \{0, 1\}$ where $p_{i,j} = 1$ determines that the execution of task i must be finished before the execution of task j can start. The synchronization constraints $s_{i,j} \in \{0, 1\}$ are given by the set $\mathcal{S} = \{s_{i,j} \,|\, i, j \in \mathcal{N}, i \neq j\}$. They constrain the execution of tasks i and j, for which $s_{i,j} = 1$ holds, to start simultaneously. An example of a heterogeneous, time-extended MRTA problem instance with precedence and synchronization constraints is depicted in Figure 3.1.

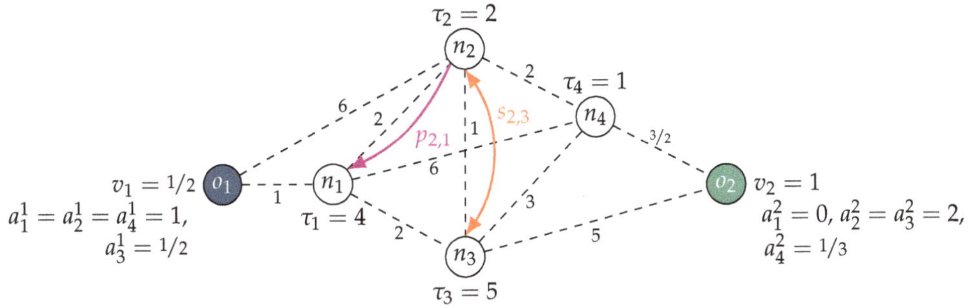

Figure 3.1: Example of a heterogeneous, time-extended MRTA problem instance $\mathcal{I}$ with precedence and synchronization constraints. It contains two agents and four tasks, i. e. $V = \{o_1, o_2, n_1, \ldots, n_4\}$. The agents differ w. r. t. their velocities and their capabilities for the four tasks, which are given next to the respective depot. The basic task durations are indicated next to the task nodes. The basic task durations together with the agents' capabilities determine the duration needed by an agent to perform a task. The dashed lines indicate the distances $\delta(i, j)$, $i, j \in V$, between the respective nodes. The graph is complete, but for better ascertainability only some distances are depicted. The distances together with the agents' velocities determine the agent-dependent edge costs. The active precedence constraint $p_{2,1} = 1$ depicted by a purple arrow constrains task n_1 not to start before task n_2 is finished. The active synchronization constraint $s_{2,3} = s_{3,2} = 1$ depicted by an orange arrow requires tasks n_2 and n_3 to start simultaneously. All other constraints are set to $p_{i,j} = s_{i,j} = 0$.

A solution $\vec{X}_\mathcal{I}$ to a time-extended MRTA problem instance $\mathcal{I}$ contains the routing information $X_\mathcal{I} = \left\{ x_{i,j}^m \middle| \, i, j \in V, m \in \mathcal{K} \right\}$, $x_{i,j}^m \in \{0, 1\}$, together with the timing information, i. e. $\vec{X}_\mathcal{I} = X_\mathcal{I} \cup \{t_i | \, i \in V\}$, $t_i \in \mathbb{R}_{\geq 0}$. The binary decision variables $x_{i,j}^m$ are equal to one if the edge between vertex i and vertex j lies on the route of agent m and zero otherwise. For the task vertices $j \in \mathcal{N} \subset V$, the real valued decision variables t_j denote the starting time of task j. For the depots $o_m \in \mathcal{O} \subset V$, the variable t_{o_m} denotes the time agent $m \in \mathcal{K}$ leaves its depot.

The optimization objective is to determine a task allocation and a task scheduling such that the solution respects all constraints and minimizes the weighted sum of task execution, transition and waiting times. With these definitions in place, the following 3-

index vehicle flow formulation (see Section A.1.1) represents the heterogeneous, time-extended MRTA problem with precedence and synchronization constraints:

Problem 3.1 (MRTA optimization problem)
For a given problem instance $\mathcal{I}$, the optimization problem to be solved is given by

$$\min_{\vec{X}_\mathcal{I}} J_\mathcal{I} = \min_{\vec{X}_\mathcal{I}} \Bigg\{ \underbrace{\sum_{m\in\mathcal{K}} \sum_{i\in V} \sum_{j\in V} d_{i,j}^m x_{i,j}^m}_{:=c_\mathcal{I}^t(\vec{X}_\mathcal{I})} + \gamma \underbrace{\sum_{m\in\mathcal{K}} \sum_{i\in\mathcal{N}} \left(d_i^m \sum_{j\in V} x_{i,j}^m \right)}_{:=c_\mathcal{I}^e(\vec{X}_\mathcal{I})} + \epsilon \underbrace{\sum_{m\in\mathcal{K}} \sum_{i\in V} \sum_{j\in V} x_{i,j}^m \left(t_j - t_i - d_i^m - d_{i,j}^m \right)}_{:=c_\mathcal{I}^w(\vec{X}_\mathcal{I})} \Bigg\} \tag{3.1}$$

subject to

$$\sum_{m\in\mathcal{K}} \sum_{i\in V} x_{i,j}^m = 1, \qquad \forall j \in V \tag{3.2a}$$
$$\sum_{i\in V} x_{i,o_m}^m = 1, \qquad \forall m \in \mathcal{K} \tag{3.2b}$$
$$\sum_{m\in\mathcal{K}} x_{i,i}^m = 0, \qquad \forall i \in \mathcal{N} \tag{3.2c}$$
$$\sum_{i\in V} x_{i,j}^m = \sum_{i\in V} x_{j,i}^m, \qquad \forall j \in V, m \in \mathcal{K} \tag{3.2d}$$
$$\sum_{m\in\mathcal{K}} \sum_{i\in V} x_{i,j}^m a_j^m > 0, \qquad \forall j \in \mathcal{N} \tag{3.2e}$$
$$x_{i,j}^m (t_i + d_i^m + d_{i,j}^m - t_j) \leq 0, \qquad \forall i \in V, j \in \mathcal{N}, m \in \mathcal{K} \tag{3.2f}$$
$$p_{i,j} \left(t_i + \sum_{m\in\mathcal{K}} \sum_{l\in V} \left(x_{i,l}^m d_{i,j}^m \right) - t_j \right) \leq 0, \qquad \forall i,j \in \mathcal{N} \tag{3.2g}$$
$$s_{i,j} \left(t_i - t_j \right) = 0, \qquad \forall i,j \in \mathcal{N} \tag{3.2h}$$
$$x_{i,j}^m \in \{0,1\}, \qquad \forall i,j \in V, m \in \mathcal{K} \tag{3.2i}$$
$$t_i \in \mathbb{R}_{\geq 0} \qquad \forall i \in V. \tag{3.2j}$$

According to (3.1), the objective function $J_\mathcal{I}$ comprises the sum of all transitioning times $c_\mathcal{I}^t(\vec{X}_\mathcal{I})$, the sum of all agent-dependent task execution times $c_\mathcal{I}^e(\vec{X}_\mathcal{I})$ and the sum of waiting times $c_\mathcal{I}^w(\vec{X}_\mathcal{I})$ associated with a solution $\vec{X}_\mathcal{I}$. This objective function extends the objective of the TSP, which aims at minimizing transitioning times [Dav10]. In addition to the overall transitioning times[22], also agent-dependent task execution times as well as waiting times are associated with a solution. Therefore, also the objectives of minimizing the overall task execution times $c^e_{\mathcal{I}}(\vec{X}_{\mathcal{I}})$ and the sum of waiting times $c^w_{\mathcal{I}}(\vec{X}_{\mathcal{I}})$ are considered in (3.1). The parameters $\gamma \in \mathbb{R}_{\geq 0}$ and $\epsilon \in \mathbb{R}_{\geq 0}$ can be used to weight the influence of the execution and waiting times on the objective function value relatively to the influence of the transitioning times.[23] For example, this can be useful if waiting times are associated with less energy consumption compared to transition or task execution times. Constraint (3.2a) ensures that each vertex is visited exactly once, constraint (3.2b) forces the routes of all agents to finally transition into their depot, constraint (3.2c) inhibits loops and constraint (3.2d) ensures that the agents' routes leave all vertices they visit. Due to constraint (3.2e), tasks must only be assigned to agents that have the capability to perform them. Constraint (3.2f) ensures the consistency of starting times associated with the solution's routing and constraints (3.2g) and (3.2h) ensure the fulfillment of the precedence and synchronization constraints, respectively. Binary decision routing variables are ensured by constraint (3.2i) and non-negative real valued starting times are enforced by constraint (3.2j).

Remark. *A linearization of Problem 3.1 is possible. Since this is not in the focus of this thesis on reoptimization, the linearization approach can be found in Appendix A.2.*

Remark. *The applied model allows for the consideration of different transition velocities of the agents. However, edge-individual differences in the transition velocities are not considered explicitely. These could be caused for example by inclines on single edges or by different driving surfaces that the agents have different competences to traverse. An inclusion of edge-individual differences in transition velocities into the model is possible by defining agent-dependent distances δ^m, $\forall m \in \mathcal{K}$. However, general metric distances are applied in Problem 3.1, since they are necessary for some of the following analyses.*[24]

A globally optimal solution to a problem instance $\mathcal{I}$ and the associated globally minimal value of the objective function are denoted by $\vec{X}^*_{\mathcal{I}} := \arg\min_{\vec{X}_{\mathcal{I}}} J_{\mathcal{I}}(\vec{X}_{\mathcal{I}})$ and $J^*_{\mathcal{I}} := J_{\mathcal{I}}(\vec{X}^*_{\mathcal{I}})$, respectively.

---

22 In contrast to the TSP, the transitioning times are agent-dependent in the heterogeneous, time-extended MRTA problem with precedence and synchronization constraints. This is considered by the agent-dependent edge costs $d^m_{i,j}$ in the transitioning times $c^t_{\mathcal{I}}(\vec{X}_{\mathcal{I}})$ of (3.1).

23 Without loss of generality, the weighting of the transitioning times in the objective function (3.1) is set to one. This is motivated by the MRTA problem being a generalization of the TSP and the TSP only minimizing the transitioning times. The guarantees given in the following chapters are also valid for transitioning time weightings in $\mathbb{R}_{>0}$.

24 The inclusion agent-dependent distances can be done without influencing the results on the feasibility of the proposed approaches, which are presented in the following sections. Furthermore, the guarantees given in the form of upper bounds on the resulting approximation ratio in the following Theorems 3.1 and 3.4 are not affected by this model choice. If agent-individual distances are applied such that for each agent $m \in \mathcal{K}$, the individual distances δ^m are metric, moreover also Theorem 3.2 and Lemma 3.9 remain valid. For the performance guarantees given in Theorem 3.3, Corollary 3.2 and Theorem 3.5 the general, uniform metric distances are necessary.

For any solution $\vec{X}_{\mathcal{I}}$ to be executable in praxis, constraints (3.2) must be fulfilled. A respective criterion to verify a solution's feasibility is given in the following section.

3.1.2 Feasibility Criterion

A criterion for the validation of a solution's feasibility is introduced in this section. The basics for these results have been established in the master thesis of Meyer [Mey19] supervised by the author and have been published at an IEEE conference by the author [BMIH20].

As depicted in Figure 3.2, the derivation of the feasibility criterion comprises three steps. First, the constraints (3.2) necessary to be fulfilled by any feasible solution are aggregated and restated in Lemma 3.1. Based on this reformulation, Lemma 3.2 gives an easy to verify criterion for the feasibility of a solution's routing $X_{\mathcal{I}}$. Finally, it is guaranteed by Lemma 3.3 that for any feasible routing $X_{\mathcal{I}}$, feasible timing information can be determined to yield a solutions $\vec{X}_{\mathcal{I}}$ that is feasible according to Lemma 3.1.

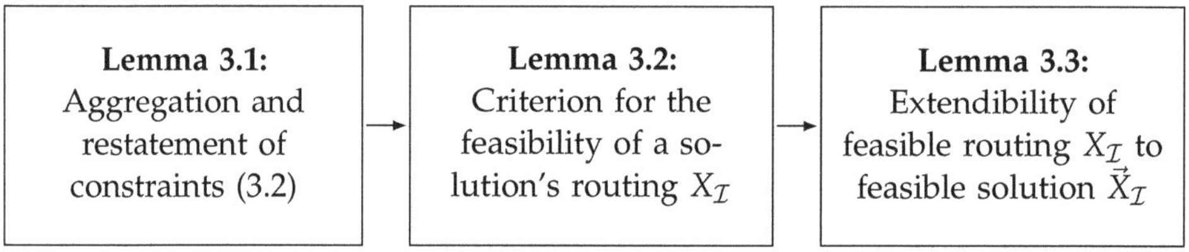

Figure 3.2: Structure of Section 3.1.2.

According to Problem 3.1, a solution $\vec{X}_{\mathcal{I}}$ is feasible for a problem instance $\mathcal{I}$ by definition, if all constraints (3.2a) to (3.2j) are fulfilled. An equivalent definition for the feasibility of a solution $\vec{X}_{\mathcal{I}}$ w. r. t. a problem instance $\mathcal{I}$ is given by the following lemma.

Lemma 3.1 (Feasibility of a solution)
A solution $\vec{X}_{\mathcal{I}}$ to a problem instance $\mathcal{I}$ according to Problem 3.1 is feasible if and only if

L3.1.1 *The agent $m \in \mathcal{K}$ assigned to any task $i \in \mathcal{N}$ by the solution $\vec{X}_{\mathcal{I}}$ is capable of its execution.*

L3.1.2 *The routing without the information of agents driving into a depot, i.e. $X_{\mathcal{I}} \setminus \left\{ x^m_{i,o_m} | i \in V, m \in \mathcal{K} \right\}$, represents a topological order and all agents $m \in \mathcal{K}$ start and end in their route in their own depot $o_m \in \mathcal{O}$.*

L3.1.3 *The positive valued, finite timing information is consistent with the routing.*

L3.1.4 *The precedence and synchronization constraints $p_{i,j} \in \mathcal{P}$ and $s_{i,j} \in \mathcal{S}$*

a) are consistent with one another and

b) are fulfilled by the solution $\vec{X}_{\mathcal{I}}$.

Proof:
Aspect L3.1.1 guarantees that constraint (3.2e) is fulfilled. Constraints (3.2a), (3.2b) (3.2c) and (3.2d) are ensured by Aspect L3.1.2. Furthermore, Aspect L3.1.3 ensures constraints (3.2f) and (3.2j) and Aspect L3.1.4.b guarantees constraints (3.2g) and (3.2h) to be fulfilled. Aspect L3.1.4.a is an additional requirement on the problem instance $\mathcal{I}$, which does not introduce any limitations since it is a necessary requirement for a problem instance to have feasible solutions. Thus, all constraints (3.2a) to (3.2j) are fulfilled if a solution $\vec{X}_{\mathcal{I}}$ fulfills Aspects L3.1.1 to L3.1.4. □

The fulfillment of Aspects L3.1.1 and L3.1.2 is determined solely by the routing $X_{\mathcal{I}}$ of a solution $\vec{X}_{\mathcal{I}}$. The routing furthermore determines, whether L3.1.4.b can be fulfilled. In the following, a criterion is derived for efficient verification of the feasibility of a routing $X_{\mathcal{I}}$ for a problem instance $\mathcal{I}$. It is furthermore proven that feasible timing information and thus a feasible solution $\vec{X}_{\mathcal{I}}$ can always be found if the routing $X_{\mathcal{I}}$ is feasible.

For the criterion for efficient verification of the feasibility of the routing $X_{\mathcal{I}}$ for a problem instance $\mathcal{I}$, a graph-based routing and constraint representation is established, the directed *constraint-solution graph (CSG)* $\tilde{G} = (\tilde{V}, \tilde{A})$. It allows for an efficient check of the routing feasibility. A CSG is generated by combining a given problem instance $\mathcal{I}$ and a corresponding routing $X_{\mathcal{I}}$. To establish the CSG, *synchronized vertices* $\tilde{v}$ are introduced which can be derived directly from a given problem instance $\mathcal{I}$. The difference of the synchronized vertices compared to the previously used vertices is that they combine all tasks connected by the synchronization constraints $s_{i,j} \in \mathcal{S}$, i. e.

$$\tilde{v}_i = \begin{cases} i & \text{if } (i \in \mathcal{O}) \vee \left[i \in \mathcal{N} \wedge \right. \\ & \quad \left. \nexists j \in \mathcal{N} : (s_{i,j} = 1 \vee s_{j,i} = 1) \right] \\ \{ j \in \mathcal{N} \,|\, j = i \vee s_{i,j} = 1 \vee s_{j,i} = 1 \} & \text{otherwise.} \end{cases} \tag{3.3}$$

Thus, each depot $o \in \mathcal{O}$ corresponds to a synchronized vertex $\tilde{v}_o$. Tasks are differentiated according to whether they are constrained by a synchronization constraint or not. For each task $i \in \mathcal{N}$ without any synchronization constraint, i. e. $s_{i,j} = s_{j,i} = 0 \; \forall \; j \in \mathcal{N}$, the synchronized vertex only contains the task itself, i. e. $\tilde{v}_i = i$. For any two tasks i and j combined by a synchronization constraint, i. e. $s_{i,j} = 1 \vee s_{j,i} = 1$, the synchronized vertices are equal, i. e. $\tilde{v}_i \equiv \tilde{v}_j$, and contain all tasks that i and j must be synchronized with.

Using the synchronized vertices according to (3.3), the constraint-solution graph is defined as follows.

Definition 3.1 (Constraint-solution graph (CSG))
A constraint-solution graph $\tilde{G}_{\mathcal{I},X_\mathcal{I}} = (\tilde{V}_\mathcal{I}, \tilde{A}_{\mathcal{I},X_\mathcal{I}})$ belongs to a problem instance $\mathcal{I} = \{\mathcal{N}, \mathcal{K}, \mathcal{O}, \mathcal{V}, \mathcal{A}, \mathcal{T}, \mathcal{D}, \mathcal{P}, \mathcal{S}\}$ and a corresponding routing $X_\mathcal{I}$. Out of all synchronized vertices $\tilde{v}$ of the problem instance $\mathcal{I}$ as given according to (3.3) a subset defines the set of vertices $\tilde{V}_\mathcal{I}$ of the CSG. The subset is chosen such that overall all tasks and all depots and are included exactly once in $\tilde{V}_\mathcal{I}$, i. e.

$$\bigcup_{\tilde{v} \in \tilde{V}} \tilde{v} = V \text{ and } \bigcap_{\tilde{v} \in \tilde{V}} \tilde{v} = \emptyset.$$

The arcs $\tilde{A}_{\mathcal{I},X_\mathcal{I}}$ of the CSG contain arcs of two different types. The routing arcs $\tilde{A}_{X_\mathcal{I}}$ contain all transitions done by any agent within the routing $X_\mathcal{I}$ towards tasks, i. e.

$$\tilde{A}_{X_\mathcal{I}} = \left\{ (f,g) \,\middle|\, \exists i \in \tilde{v}_f, j \in (\tilde{v}_g \cup \mathcal{N}) : \sum_{m \in \mathcal{K}} x_{i,j}^m = 1 \right\}.$$

Arcs entering depot vertices are thus not included in the routing arcs. Together with the precedence arcs $\tilde{A}_\mathcal{P}$ that are defined by all active precedence constraints, i. e.

$$\tilde{A}_\mathcal{P} = \left\{ (f,g) \,\middle|\, \exists i \in \tilde{v}_f, j \in \tilde{v}_g : p_{i,j} = 1 \right\},$$

the union of the solution and precedence arcs defines the set of arcs of the CSG, i. e.

$$\tilde{A}_{\mathcal{I},X_\mathcal{I}} = \tilde{A}_{X_\mathcal{I}} \cup \tilde{A}_\mathcal{P}.$$

An example of a CSG is depicted in Figure 3.3. Figure 3.3a depicts a problem instance $\mathcal{I}$ with two agents and six tasks. The active precedence constraints $p_{2,1} = 1$ and $p_{5,3} = 1$ are depicted by purple arrows. All other precedence constraints are equal to zero. An orange double arrow indicates the only active synchronization constraint $s_{3,4} = s_{4,3} = 1$. A possible routing $X_{\mathcal{I}}$ fulfilling the routing requirements according to L3.1.2 and L3.1.4 of Lemma 3.1 is depicted in Figure 3.3b. Routing information corresponding to agent k_1 is given in blue and the one for agent k_2 is given in green, respectively. The corresponding CSG is shown in Figure 3.3c.

The following Lemma 3.2 defines a criterion that uses the CSG to verify the feasibility of a routing according to Lemma 3.1.

Lemma 3.2 (Routing feasibility criterion)
A routing $X_{\mathcal{I}}$ is feasible for a problem instance $\mathcal{I}$ if and only if

L3.2.1 $\sum_{m \in \mathcal{K}} \sum_{j \in V} x_{i,j}^m a_i^m > 0$ *holds for all tasks* $i \in \mathcal{N}$,

L3.2.2 the corresponding CSG $\tilde{G}_{\mathcal{I},X_{\mathcal{I}}}$ *is acyclic and*

L3.2.3 The routes of all agents $m \in \mathcal{K}$ *start and end in their individual depot* $o_m \in \mathcal{O}$.

Proof:
By definition, L3.2.1, i. e. $\sum_{m \in \mathcal{K}} \sum_{j \in V} x_{i,j}^m a_i^m > 0 \; \forall \; i \in \mathcal{N}$, is equivalent to Aspect L3.1.1 of Lemma 3.1. The fulfillment of the feasibility Aspect L3.1.1 is thus defined solely by the routing information $X_{\mathcal{I}}$.

It holds that a topological ordering of a directed graph is possible if and only if the graph is acyclic [SW12, Chapter 4.2]. Therefore, the routing $X_{\mathcal{I}}$ represents a topological ordering, if and only if $\tilde{G}_{X_{\mathcal{I}}} = (\tilde{V}_{\mathcal{I}}, \tilde{A}_{X_{\mathcal{I}}})$ is acyclic. For the same reason, the set of precedence constaints $\mathcal{P}$ together with the set of synchronization constraints $\mathcal{S}$ fulfill L3.1.4.a if and only if $\tilde{G}_{\mathcal{P}} = (\tilde{V}_{\mathcal{I}}, \tilde{A}_{\mathcal{P}})$ is acyclic. In order for the routing $X_{\mathcal{I}}$ to be feasible, it must not contradict the precedence and synchronization constraints, i. e. the topological orders of $\tilde{G}_{X_{\mathcal{I}}} = (\tilde{V}_{\mathcal{I}}, \tilde{A}_{X_{\mathcal{I}}})$ and of $\tilde{G}_{\mathcal{P}} = (\tilde{V}_{\mathcal{I}}, \tilde{A}_{\mathcal{P}})$ must not contradict each other. This is the case if $\tilde{G}_{\mathcal{I},X_{\mathcal{I}}} = (\tilde{V}_{\mathcal{I}}, \tilde{A}_{\mathcal{I},X_{\mathcal{I}}})$ is acyclic. Consequently, L3.2.2 ensures the routing $X_{\mathcal{I}}$ to represent a topological ordering, the precedence and synchronization constraints to be consistent with one another and the routing to be such that it does not contradict the precedence and synchronization constraints. Together with L3.2.3, Aspect L3.1.2 of Lemma 3.1 is also ensured to be fulfilled. □

Thus, given a problem instance $\mathcal{I}$ and any routing $X_{\mathcal{I}}$, Lemma 3.2 can be used to verify its feasibility. Hence, by reformulating a problem instance $\mathcal{I}$ and a corrresponding routing$X_{\mathcal{I}}$ using the related CSG, the feasibility verification of the routing mainly reduces to the well studied problem of acyclicity checks within graphs. Any cycle search or acyclicity check (e. g. [Kah62, BR02]) can be applied.

Lemma 3.3 states that for any feasible routing a complete feasible solution is always attainable.

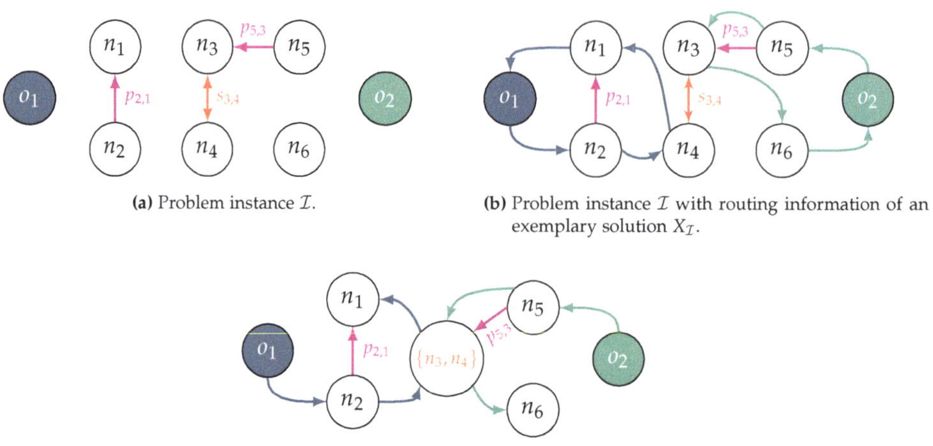

(a) Problem instance $\mathcal{I}$.

(b) Problem instance $\mathcal{I}$ with routing information of an exemplary solution $X_{\mathcal{I}}$.

(c) Corresponding CSG $\tilde{G}_{\mathcal{I},X_{\mathcal{I}}}$.

Figure 3.3: Example of a constraint-solution graph.

Lemma 3.3 (Solution feasibility for feasible routing)
For any routing $X_{\mathcal{I}}$ that is feasible for a problem instance $\mathcal{I}$ according to Lemma 3.2, timing information can be determinded such that the resulting solution $\vec{X}_{\mathcal{I}}$ is feasible according to Lemma 3.1. The timing information $\{t_i|\forall i \in V\}$ feasible for routing $X_{\mathcal{I}}$ fulfills the following constraints:

$$x^m_{i,j}(t_i + d^m_i + d^m_{i,j} - t_j) \leq 0, \qquad \forall i \in V, j \in \mathcal{N}, m \in \mathcal{K} \tag{3.4a}$$

$$p_{i,j}\left(t_i + \sum_{m\in\mathcal{K}}\sum_{l\in V}\left(x^m_{i,l}d^m_i\right) - t_j\right) \leq 0, \quad \forall i,j \in \mathcal{N} \tag{3.4b}$$

$$s_{i,j}\left(t_i - t_j\right) = 0, \qquad \forall i,j \in \mathcal{N} \tag{3.4c}$$

$$t_i \in \mathbb{R}_{\geq 0}, \qquad \forall i \in V \tag{3.4d}$$

$$x^m_{i,j} \in X_{\mathcal{I}}, \qquad \forall i \in V, j \in \mathcal{N}, m \in \mathcal{K}. \tag{3.4e}$$

Proof:
Given the feasibility of a routing $X_{\mathcal{I}}$ and the acyclicity of the corresponding CSG $\tilde{G}_{\mathcal{I},X_{\mathcal{I}}}$, a topological sorting of the vertices defining a partial order is possible. Using the partial order, starting times t_i, $\forall i \in V$, respecting the routing $X_{\mathcal{I}}$ information can be defined recursively. Constraints (3.4a) to (3.4e) ensure the tasks' durations and transitioning times as well as the precedence and synchronization constraints to be considered and thus the resulting solution to be feasible for the corresponding problem instance $\mathcal{I}$. Constraint (3.4a) ensures the starting times to be consistent with the routing information since no vertex j can be started before the vertex i directly precedent according to routing $X_{\mathcal{I}}$ is finished and the respective agent has traversed from i to j and constraint (3.4e) enforces that the routing $X_{\mathcal{I}}$ is considered. Together with the

feasibility of the routing $X_\mathcal{I}$, this ensures Aspect L3.1.3 of Lemma 3.1 to be fulfilled. Constraints (3.4b) and (3.4c) require the precedence and synchronization constraints to be fulfilled which ensures Aspect L3.1.4.b of Lemma 3.1. Constraint (3.4d) ensures the starting times to be within the domain definition. Together with the feasibility of the routing $X_\mathcal{I}$, feasibility according to Lemma 3.1 is ensured. □

When looking at the objective function (3.1), the transistion costs $c^t_\mathcal{I}(\vec{X}_\mathcal{I})$ and the execution costs $c^e_\mathcal{I}(\vec{X}_\mathcal{I})$ are fully defined by the routing information $X_\mathcal{I}$ of the complete solution $\vec{X}_\mathcal{I}$. The waiting costs however, are influenced by all components of the complete solution $\vec{X}_\mathcal{I}$. Thus, a complete solution $\vec{X}_\mathcal{I}$ that optimizes the objective function for a given, feasible routing $X_\mathcal{I}$, optimizes the resulting waiting times w. r. t. the routing. In order to derive the corresponding timing information, the following optimization problem needs to be solved:

$$\min_{\{t_i|\forall i \in V\}} \left\{ \sum_{m\in\mathcal{K}} \sum_{i\in V} \sum_{j\in V} x^m_{i,j} \left(t_j - t_i - d^m_i - d^m_{i,j} \right) \right\} \tag{3.5}$$

subject to constraints (3.4).

Corollary 3.1
A solution $\vec{X}_\mathcal{I}$ to a problem instance $\mathcal{I}$ that uses a feasible routing information $X_\mathcal{I}$ and defines the timing information according to the optimization problem given by (3.5), is feasible according to Lemma 3.1.

So far, the heterogeneous, time-extended MRTA optimization problem with precedence and synchronization constraints is defined and criteria to verify a solution's feasibility are introduced. In the following section, a generalized version of the corresponding reoptimization problem is given.

3.1.3 Generalized Time-Extended MRTA Reoptimization Problem

In this section, a general definition of the reoptimization problem corresponding to the previously defined heterogeneous, time-extended MRTA optimization problem with precedence and synchronization constraints (see Problem 3.1) is given. It assumes an initial problem instance $\mathcal{I}$ and a globally optimal solution $\vec{X}^*_\mathcal{I}$ to the initial problem instance to be given. A modified problem instance $\mathcal{I}_{\text{mod}}$, that differs from the initial problem instance $\mathcal{I}$ in a defined manner shall be solved using the knowledge of the initial instance, its globally optimal solution and the modification applied to the inital problem instance (see Figure 2.2). Problem 3.2 defines the heterogeneous, time-extended MRTA reoptimization problem with precedence and synchronization constraints.

Problem 3.2 (MRTA reoptimization problem)
Let an initial MRTA problem instance $\mathcal{I} = \{\mathcal{N}, \mathcal{K}, \mathcal{O}, \mathcal{V}, \mathcal{A}, \mathcal{T}, \mathcal{D}, \mathcal{P}, \mathcal{S}\}$ according to Problem 3.1 and a globally optimal solution $\vec{X}^*_{\mathcal{I}}$, that is feasible according to Lemma 3.1, be given. Let furthermore a modified MRTA problem instance $\mathcal{I}_{mod}$ according to Problem 3.1 be given, that differs from the initial problem instance $\mathcal{I}$ by a defined modification.
The MRTA reoptimization problem aims at finding a solution $\vec{X}^{reo}_{\mathcal{I}_{mod}}$ to the modified problem instance $\mathcal{I}_{mod}$ by making use of the knowledge of the applied modification and the information contained in the globally optimal solution $\vec{X}^*_{\mathcal{I}}$ to the initial problem instance $\mathcal{I}$ such that the resulting modified solution $\vec{X}^{reo}_{\mathcal{I}_{mod}}$ fulfills all constraints according to (3.2) *of the modified problem instance $\mathcal{I}_{mod}$ and optimizes objective function* (3.1).

The modification applied to the initial problem instance strongly influences the heuristic solution approaches that can be applied to solve the corresponding reoptimization problem. In this thesis, reoptimization heuristics for ten relevant modifications as listed in Table 3.1 are introduced and analyzed. They include important modifications such as task insertion and task deletion that have already been investigated for the TSP (see Table 2.4 in Section 2.5.2) and are extended by modification possibilities that arise from the augmented problem statement (in comparison to the TSP) of heterogeneous, time-extended MRTA problems with precedence and synchonization constraints.

An overview of the relations between the applied reoptimization heuristics is depicted in Figure 3.4. The cheapest maximum insertion cost heuristic (CMI), the task deletion heuristic (TDH), and the initial solution approach (INI) are defined independently of each other. The delete-insert heuristic (DIH) results from the combination of an extension of the CMI called extended CMI (eCMI) and the TDH. Finally, the extended DIH (eDIH) is an extension of the DIH.

In the following section, the task insertion reoptimization problem is analyzed w. r. t. performance guarantees of applied solution heuristics.

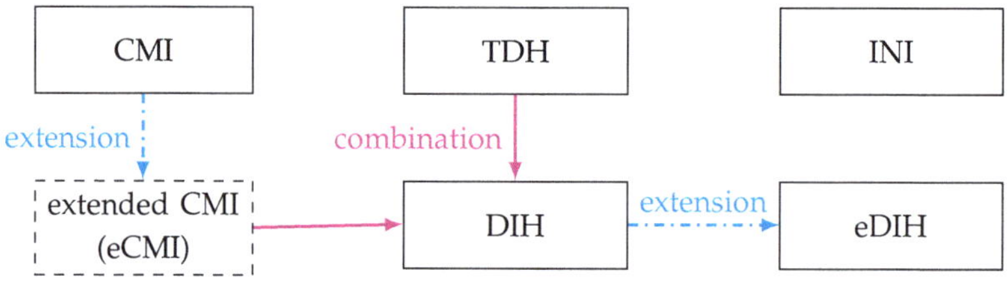

Figure 3.4: Overview of the relation between the applied reoptimization heuristics.

Table 3.1: Overview of analyzed modifications and applied heuristics.

Modification	Applied reoptimization heuristics
task insertion	cheapest maximum insertion cost heuristic (CMI)
task deletion	task deletion heuristic (TDH)
task position variation	initial solution approach (INI) and delete-insert heuristic (DIH)
task duration variation	initial solution approach (INI) and delete-insert heuristic (DIH)
agent capability variation	initial solution approach (INI)
agent velocity variation	initial solution approach (INI)
precedence constraint insertion	extended delete-insert heuristic (eDIH)
synchronization constraint insertion	extended delete-insert heuristic (eDIH)
precedence constraint deletion	initial solution approach (INI)
synchronization constraint deletion	initial solution approach (INI)

3.2 Task Insertion

In this section, the task insertion reoptimization problem corresponding to the time-extended MRTA optimization problem defined in Problem 3.1 is introduced and analyzed. First, two assumptions on the relation between the objective function value of the initial and the modified problem instance as well as on the objective function increment resulting from the task insertion are introduced. Upon these, it is proven in this thesis that for any reoptimization approach that fulfills these two assumptions, a performance guarantee in the form of an upper bound of $\alpha \leq 2$ on the approximation ratio can be given. Subsequently, the *cheapest maximum insertion cost heuristic (CMI)* is introduced, and it is shown that the CMI always fulfills the aforementioned assumptions. Furthermore, it is proven that solutions found by the CMI are always feasible. To give even smaller upper bounds on the approximation ratios, three problem configurations are introduced, differing in whether heterogeneous groups of agents and temporal task constraints are considered. For temporally unconstrained problem instances smaller upper bounds on the approximation ratios are derived, the smallest one being $\alpha \leq 3/2$ for homogeneous groups of agents. The bounds are proven to be tight, meaning that no smaller upper bounds for the CMI exist. The following more detailed presentation of these results is based on a journal publication of the author [BKH$^+$24].

3.2.1 Task Insertion Reoptimization Problem

The MRTA task insertion modification extends an initial problem instance $\mathcal{I}$ by an additional task n_{N+1} which yields the modified problem instance $\mathcal{I}^+$. The corresponding MRTA task insertion reoptimization problem aims at finding a feasible solution to the modified problem instance $\mathcal{I}^+$ by making use of the knowledge of a globally optimal solution $\vec{X}^*_{\mathcal{I}}$ to the initial problem instance $\mathcal{I}$. A comprehensive formal definition is given in Problem 3.3.

Problem 3.3 (MRTA task insertion reoptimization problem)
The MRTA task insertion reoptimization problem is an MRTA reoptimization problem according to Problem 3.2, where the modified problem instance $\mathcal{I}^+$ is given by $\mathcal{I}^+ = \{\mathcal{N}^+, \mathcal{K}, \mathcal{O}, \mathcal{V}, \mathcal{A}^+, \mathcal{T}^+, \mathcal{D}^+, \mathcal{P}^+, \mathcal{S}^+\}$ with the modified sets

- $\mathcal{N}^+ = \mathcal{N} \cup \{n_{N+1}\}$,
- $\mathcal{A}^+ = \mathcal{A} \cup \{a^m_{N+1} \,|\forall m \in \mathcal{K}\}$,
- $\mathcal{T}^+ = \mathcal{T} \cup \{\tau_{N+1}\}$,
- $\mathcal{D}^+ = \mathcal{D} \cup \{\delta(i, n_{N+1}), \delta(n_{N+1}, i) \,|\forall i \in V\}$,
- $\mathcal{P}^+ = \mathcal{P} \cup \{p_{i,N+1} = p_{N+1,i} = 0 \,|\forall i \in \mathcal{N}\}$ *and*
- $\mathcal{S}^+ = \mathcal{S} \cup \{s_{i,N+1} = s_{N+1,i} = 0 \,|\forall i \in \mathcal{N}\}$.

Thus, the modified problem instance $\mathcal{I}^+$ contains an additional temporarily unconstrained task n_{N+1} compared to the initial problem instance $\mathcal{I}$.

On the basis of this definition of the MRTA task insertion reoptimization problem as given in Problem 3.3, the following section derives a guarantee on the solution quality that holds independently of the applied reoptimization approach.

3.2.2 Reoptimization Approach Independent Guarantee on the Approximation Ratio

In this section, it is shown that any solution approach to solve the MRTA task insertion reoptimization problem (Problem 3.3), under certain assumptions, cannot yield a reoptimization ratio greater than 2.

The first assumption considers the relation the of optima of the objective function values of the initial and the modified problem instance of the MRTA reoptimization problem.

Assumption 3.1. *Let*

$$J^*_{\mathcal{I}^+} \geq J^*_{\mathcal{I}} \tag{3.6}$$

be fulfilled for the modified problem instance $\mathcal{I}^+$ and the initial problem instance $\mathcal{I}$ of the MRTA task insertion reoptimization Problem 3.3.

The following lemma provides two conditions that ensure Assumption 3.1 to be fulfilled.

Lemma 3.4

If

1. *the weights γ, ϵ of the objective function* (3.1) *of the underlying optimization Problem 3.1 of the MRTA task insertion reoptimization Problem 3.3 fulfill*

$$\epsilon \leq \min\{1, \gamma\} \tag{3.7}$$

or

2. *within the MRTA task insertion reoptimization problem instance (see Problem 3.3) neither precedence nor synchronization constraints must be considered,*

Assumption 3.1 is fulfilled.

Proof:

The modified problem instance $\mathcal{I}^+$ differs from the initial problem instance $\mathcal{I}$ only in comprising the additional task n_{N+1}. The objective function 3.1 consists of the weighted sum of transition times c^t, task execution times c^e, and waiting times c^w. By introducing the additional task n_{N+1}, the transition times c^t and task execution times c^e can only increase or remain constant. The waiting times c^w can both increase or decrease. However, the decrease of waiting times is limited to be of the same magnitude as the sum of the increases in transition and task execution times. This is the case, if the additional transition and task execution times replace former waiting times. By denoting the differences in transition, task execution and waiting times as $\Delta_{c^t} \in \mathbb{R}_{\geq 0}$, $\Delta_{c^e} \in \mathbb{R}_{\geq 0}$ and $\Delta_{c^w} \in \mathbb{R}$, $\Delta_{c^w} \geq -(\Delta_{c^t} + \Delta_{c^e})$, respectively, it thus holds for the difference in optimal objecitve function values

$$\begin{aligned} J^*_{\mathcal{I}^+} - J^*_{\mathcal{I}} &= \Delta_{c^t} + \gamma\Delta_{c^e} + \epsilon\Delta_{c^w} \\ &\geq (1-\epsilon)\Delta_{c^t} + (\gamma - \epsilon)\Delta_{c^e} \end{aligned} \tag{3.8}$$

Hence, condition (3.7) is sufficient for the fulfillment of Assumption 3.1.

If neither precedence nor synchronization constraints are contained in the MRTA task insertion reoptimization problem instance, no waiting times occur in the optimal solutions $\vec{X}^*_{\mathcal{I}}$ and $\vec{X}^*_{\mathcal{I}^+}$. Consequently, it holds that $\Delta_{c^w} = 0$ and Assumption 3.1 is fulfilled independently of the choice of ϵ and γ in these cases. □

Furthermore, any MRTA task insertion reoptimization approach that uses the initial solution $\vec{X}^*_{\mathcal{I}}$ to solve the modified problem instance $\mathcal{I}^+$, yields a reoptimized solution $\vec{X}^{\text{reo}}_{\mathcal{I}^+}$ that adds an increment in the objective function $\Delta^+ \in \mathbb{R}$ to the optimal objective function value $J^*_{\mathcal{I}}$ of the initial problem instance.[25] Thus, the resulting objective

[25] Under Assumption 3.1, the increment Δ^+ in the objective function is ensured to be non-negative.

function value of the reoptimized solution $J_{\mathcal{I}^+}^{\text{reo}}$ equals the sum of the initial optimal objective function value $J_{\mathcal{I}}^*$ and the increment in the objective function Δ^+, i. e.

$$J_{\mathcal{I}^+}^{\text{reo}} = J_{\mathcal{I}}^* + \Delta^+. \tag{3.9}$$

Depending on the specific problem instance and on the reoptimization method used to solve the task insertion reoptimization problem, the increment in the objective function value Δ^+ resulting from the application of the respective reoptimization approach to a task insertion problem instance may differ. For the following analysis, the increment Δ^+ is conservatively assumed to have a value at most as high as the objective function value $J_{\mathcal{I}^+}^*$ of the optimal solution to the modified problem instance.

Assumption 3.2. *Let the increment Δ^+ in the value of the objective function resulting from solving the MRTA task insertion reoptimization problem according to Problem 3.3 be bounded above by the value of the objective function $J_{\mathcal{I}^+}^*$ of a globally optimal solution to the modified problem instance $\mathcal{I}^+$, i. e.*

$$\Delta^+ \leq J_{\mathcal{I}^+}^*. \tag{3.10}$$

For any reoptimization solution approach solving an instance of the MRTA task insertion problem such that Assumptions 3.1 and 3.2 hold, a performance guarantee can be given according to the following theorem.

Theorem 3.1

For every MRTA task insertion reoptimization approach solving an instance of Problem 3.3 such that Assumptions 3.1 and 3.2 hold, the resulting approximation ratio $\alpha = J_{\mathcal{I}^+}^{reo} / J_{\mathcal{I}^+}^*$ is bounded above by 2, i. e.

$$\alpha = \frac{J_{\mathcal{I}^+}^{reo}}{J_{\mathcal{I}^+}^*} \leq 2. \tag{3.11}$$

Proof:

Using (3.6) from Assumtion 3.1 together with (3.10) from Assumption 3.2, it follows for the approximation ratio

$$\alpha = \frac{J_{\mathcal{I}^+}^{\text{reo}}}{J_{\mathcal{I}^+}^*} = \frac{J_{\mathcal{I}}^* + \Delta^+}{J_{\mathcal{I}^+}^*} \overset{(3.10)}{\leq} \frac{J_{\mathcal{I}}^* + J_{\mathcal{I}^+}^*}{J_{\mathcal{I}^+}^*} \overset{(3.6)}{\leq} 2. \tag{3.12}$$

□

Remark. *The performance guarantee given by Theorem 3.1 in the form of upper bounds on the approximation ratio for the task insertion reoptimization problem (Problem 3.3) holds independent of the specific reoptimization approach used to solve an MRTA task insertion reoptimization problem instance $\mathcal{I}^+$, as long as Assumptions 3.1 and 3.2 are fulfilled. While the* *fulfillment of Assumption 3.2 depends on the applied reoptimization approach, the fulfillment of Assumption 3.1 is influenced by the specific task insertion reoptimization problem instance under consideration.*

According to Lemma 3.4, Assumption 3.1 is guaranteed to be fulfilled, if either the objecitve function (3.1) *weights waiting times at most as much as task execution and transitioning times, or if no waiting times can occur since no temporal constraints are to be considered within the problem instance. Thus, the choice of the weights of the objective function according to* (3.7) *together with any reoptimization approach that fulfills Assumption 3.1, guarantees that the reoptimized solution* $\vec{X}^{reo}_{\mathcal{I}^+}$ *cannot have an objective function value* $J^{reo}_{\mathcal{I}^+}$ *of more than twice the value of the objective function* $J^*_{\mathcal{I}^+}$ *of a globally optimal solution to the modified problem instance* $\mathcal{I}^+$.

Since the conditions given in Lemma 3.4 are sufficient, Assumption 3.1 still holds in many cases that do not fulfill either of the conditions. For example, if the objective function (3.1) *weights waiting times more than task execution or transition times, Assumption 3.1 is still fulfilled for each problem instance in which the additional task exectution and transition times caused by the insertion of the additional task* n_{N+1} *does not replace former waiting times. However, the upper bound on the approximation ratio of* $\alpha \leq 2$ *is not guaranteed in general for such a choice of weights* ϵ, γ *within the objective function* (3.1) *if temporal constraints can be contained within the problem instances.*

In the following, the *cheapest maximum cost insertion heuristic*, a specific reoptimization heuristic for Problem 3.3, is introduced.

3.2.3 Cheapest Maximum Insertion Cost Heuristic

In this section, a heuristic solution called the *cheapest maximum cost insertion heuristic (CMI)* to solve the MRTA task insertion reoptimization problem (Problem 3.3) is introduced. It is inspired by the "cheapest insertion heuristic" (CI) which was proposed by Archetti et al. [ABS03] for the TSP. The CI chooses to insert the new task to the initial solution such that the resulting cost increment, i. e. the increment in the value of the objective function, is minimized. While the CI is based on the determination of the exact cost increment, the CMI uses an overapproximation of the increase of the objective function value associated with inserting the new task n_{N+1} on an edge of an agent's route of the initial solution. This is due to the fact that precedence and synchronization constraints present within a time-extended MRTA problem instance can cause waiting times $c^w_{\mathcal{I}} \neq 0$ that influence the objective function (3.1). These waiting times must therefore be considered when determining the cost increment. Even though the new task n_{N+1} is not associated with any temporal constraint, the tasks following within the same route might have to consider precedence or synchronization constraints and their temporal shifts can therefore cause an increase in waiting times in the routes of other agents. In the same manner, the temporal shifts caused by the additional waiting times of these tasks might cause further additional waiting times and associated shifts in task starting times within the routes of even more agents. This effect makes the determination of exact insertion costs much more challenging which is why a less complex overestimation of the insertion costs called the *maximum insertion costs* $\Delta_{\text{max}}^{i,j,m}$ is proposed. The maximum insertion costs use the exact additional task execution and transition times and a possible overestimation of the additional waiting times resulting from inserting the additional task n_{N+1} in the route of agent m on the edge (i,j).[26]

To derive the maximum possible increase in waiting times, $\beta_j \in \{1,\ldots,K\}$ describes the maximum number of agents' routes that might be affected of a temporal shift of vertex $j \in V$ within the initial solution $\vec{X}_{\mathcal{I}}^*$. This implies that the temporal shift of vertex j might, in the worst case, cause additional waiting times of the same amount as the initial temporal shift of vertex j in the routes of $\beta_j - 1$ agents other than the one vertex j is assigned to. Consequently, the maximum insertion cost resulting from inserting task n_{N+1} between the vertices i and j in the route of agent $m \in \mathcal{K}$ is given by the additional transition and task execution times for agent m which are associated with this insertion plus the waiting times of the same amount in at most $\beta_j - 1$ routes, i. e.

$$\begin{aligned}
\Delta_{\text{max}}^{i,j,m} &= \begin{cases} \Delta_{\text{trans}}^{i,j,m} + \gamma d_{N+1}^{m} + \epsilon\left(\beta_j - 1\right)\left(\Delta_{\text{trans}}^{i,j,m} + d_{N+1}^{m}\right) & \text{if } a_{N+1}^{m} > 0 \\ \infty & \text{if } a_{N+1}^{m} = 0 \end{cases} \\
&= \begin{cases} \Delta_{\text{trans}}^{i,j,m} + \gamma \frac{\tau_{N+1}}{a_{N+1}^{m}} + \epsilon\left(\beta_j - 1\right)\left(\Delta_{\text{trans}}^{i,j,m} + \frac{\tau_{N+1}}{a_{N+1}^{m}}\right) & \text{if } a_{N+1}^{m} > 0 \\ \infty & \text{if } a_{N+1}^{m} = 0 \end{cases} \\
&\qquad \forall\,(i,j,m) : x_{i,j}^{m} \in X_{\mathcal{I}}^{*} \wedge x_{i,j}^{m} = 1
\end{aligned} \tag{3.13}$$

with

$$\begin{aligned}
\Delta_{\text{trans}}^{i,j,m} &= d_{i,N+1}^{m} + d_{N+1,j}^{m} - d_{i,j}^{m} \\
&= \frac{\delta(i,N+1)}{v_m} + \frac{\delta(N+1,j)}{v_m} - \frac{\delta(i,j)}{v_m}.
\end{aligned} \tag{3.14}$$

Using the maximum insertion costs, the CMI applied to MRTA task insertion problem instances according to Problem 3.3 is defined as given by the following algorithm.

Remark. *To determine the maximal number β_i of agents' routes within which temporal shifts of vertices may occur due to a temporal shift of vertex $i \in V$, the CSG $\tilde{G}_{\mathcal{I},X_{\mathcal{I}}^*}$ corresponding to the globally optimal solution $\vec{X}_{\mathcal{I}}^*$ of the initial problem instance $\mathcal{I}$ can be used. Given the CSG $\tilde{G}_{\mathcal{I},X_{\mathcal{I}}^*}$, β_i is for all vertices $i \in V$ given by the number of agents' depots $o_m \in \mathcal{O}$ for which*

[26] The overestimation of additional waiting times differentiates the CMI from the CI applied by Archetti [ABS03] for the TSP.

Algorithm 1 Cheapest maximum insertion cost heuristic (CMI)

1: For each edge (i, j) that is part of the route of any agent $m \in \mathcal{K}$ in the routing $X^*_{\mathcal{I}}$ of the optimal solution $\vec{X}^*_{\mathcal{I}}$ of the initial problem instance $\mathcal{I}$, i. e. for all $(i, j, m) : \left(x^m_{i,j} \in X^*_{\mathcal{I}} \wedge x^m_{i,j} = 1\right)$, calculate the maximum insertion costs $\Delta^{i,j,m}_{\max}$ according to (3.13).

2: Insert the added task n_{N+1} in the route of agent $\hat{m}$ between the vertices $\hat{i}$ and $\hat{j}$ that correspond to the finite minimum of the above determined maximum insertion costs $\Delta^{i,j,m}_{\max}$. This yields the reoptimized routing $X^{\mathrm{CMI}}_{\mathcal{I}^+}$. The resulting optimal maximum insertion cost $\Delta^*_{\max} \in \mathbb{R}_{\geq 0}$ is given by

$$\Delta^*_{\max} = \Delta^{\hat{i},\hat{j},\hat{m}}_{\max} = \min_{\left\{(i,j,m) \middle| x^m_{i,j} \in X^*_{\mathcal{I}} \wedge x^m_{i,j}=1\right\}} \Delta^{i,j,m}_{\max}, \tag{3.15}$$

3: Solve optimization problem (3.5) to determine the complete CMI solution corresponding to routing $X^{\mathrm{CMI}}_{\mathcal{I}^+}$, i. e. to complement the routing with the timing information to complete the solution $\vec{X}^{\mathrm{CMI}}_{\mathcal{I}^+}$.

there exists a path from vertex i to the corresponding depot within the directed graph $\tilde{G}_{\mathcal{I},X^*_{\mathcal{I}}}$, *i. e.*

$$\beta_i = \left|\left\{o_m \in \mathcal{O} : \exists \text{ path } (i, \ldots, o_m) \text{ in } \tilde{G}_{\mathcal{I},X^*_{\mathcal{I}}}\right\}\right|. \tag{3.16}$$

Hence, β_i *can be determined using e. g. a backwards breadth-first search within the CSG* $\tilde{G}_{\mathcal{I},X^*_{\mathcal{I}}}$ *[Koh21].*

Since the CMI chooses the edge on which to insert the new task n_{N+1} based on a potential overestimation of the actual increment in the objective function value, i. e.

$$\Delta^*_{\max} \geq \Delta^+, \tag{3.17}$$

the value of the objective function $J^{\mathrm{CMI}}_{\mathcal{I}^+} = J^*_{\mathcal{I}} + \Delta^+$ (see (3.9)) of the solution $\vec{X}^{\mathrm{CMI}}_{\mathcal{I}^+}$ generated by the CMI is bounded above by

$$J^{\mathrm{CMI}}_{\mathcal{I}^+} \leq J^*_{\mathcal{I}} + \Delta^*_{\max}. \tag{3.18}$$

In the following section, the feasibility of the solutions generated by the CMI is analyzed.

3.2.4 Feasibility of the Solutions Generated by the Cheapest Maximum Insertion Cost Heuristic

According to Lemma 3.5, if feasible solutions to the modified problem instance $\mathcal{I}^+$ exist, the CMI is ensured to yield a feasible solution.

Lemma 3.5 (Solution feasibility for CMI)

Let an MRTA task insertion reoptimization problem according to Problem 3.3 be given such that the solution set of the modified problem instance $\mathcal{I}^+$ is non-empty. The CMI (see Algorithm 1) always finds a solution $\vec{X}_{\mathcal{I}^+}^{\mathrm{CMI}}$ to the modified problem instance $\mathcal{I}^+$ that is feasible according to Lemma 3.1.

Proof:

According to the definition of Problem 3.3, the initial solution $\vec{X}_{\mathcal{I}}^*$ is feasible. Furthermore, the optimal maximum insertion cost $\Delta_{\max}^*$ must be finite (see Algorithm 1). According to the definition of the optimum maximum insertion costs (see (3.13), (3.14) and (3.15)) this is exactly the case if there exists at least one agent $m \in \mathcal{K}$ with a capability $a_{N+1}^m > 0$ to perform the inserted task n_{N+1}. The existence of at least one such agent is ensured for any modified problem instance with a non-empty solution set since otherwise constraint (3.2e) could not be fulfilled for the optimization Problem 3.1 associated with $\mathcal{I}_{\mathrm{mod}}$. Consequently, Aspect L3.2.1 of the routing feasibility criterion (Lemma 3.2) is alway fulfilled for a routing defined by the CMI. Also, since the initial solution is feasible it has an acyclic CSG. Since by Definition 3.3, the inserted task n_{N+1} is not constrained by any precedence or synchronization constraint, inserting the task n_{N+1} on any edge of the initial routing cannot introduce a cycle to the resulting CSG. Consequently, the routing generated by the CMI always fulfills L3.2.2. Furthermore, inserting task n_{N+1} on only one edge of the initial routing does not alter start and end nodes of the routing of any agent. Since the initial solution is feasible, the CMI routing is furthermore ensured to fulfill L3.2.3, i. e. the agents start and end their routes at their individual depot. Consequently, since all aspects of Lemma 3.2 are ensured to be fulfilled by the routing generated by the CMI, the respective routing is always feasible. By definition of the CMI (see Algorithm 1), the timing information to complete the CMI solution $\vec{X}_{\mathcal{I}^+}^{\mathrm{CMI}}$ is determined by solving optimization problem (3.5). According to Corollary 3.1, this approach ensures the solution $\vec{X}_{\mathcal{I}^+}^{\mathrm{CMI}}$ to be feasible. □

An analysis of the approximation ratios resulting from the application of the CMI is presented in the following.

3.2.5 Upper Bounds on the Approximation Ratio for the Cheapest Maximum Insertion Cost Heuristic

As presented above, a solution generated by the CMI (see Lemma 3.5) is ensured to be feasible according to Lemma 3.2. Extending this, performance guarantees in the form of upper bounds on the approximation ratios are given in this section.

The following lemma holds for metric distances when inserting a vertex to a graph.

Lemma 3.6

When a vertex n_{N+1} is inserted on an edge $(i,j) \in E$, it holds for metric distances that

$$\delta(i, n_{N+1}) + \delta(n_{N+1}, j) - \delta(i,j) \leq 2\delta(n_{N+1}, j). \tag{3.19}$$

Proof:
Since the distances are metric, the triangle inequality $\delta(i, n_{N+1}) \leq \delta(i,j) + \delta(n_{N+1}, j)$ holds. Together with the symmetry of distances, i. e. $\delta(n_{N+1}, j) = \delta(j, n_{N+1})$, the assertion follows:

$$\begin{aligned} \delta(i, n_{N+1}) &\leq \delta(i,j) + \delta(j, n_{N+1}) = \delta(i,j) + \delta(n_{N+1}, j) \\ \Leftrightarrow \delta(i, n_{N+1}) &+ \delta(n_{N+1}, j) - \delta(i,j) \leq 2\delta(n_{N+1}, j) \end{aligned}$$

□

Using Lemma 3.6, the following lemma shows that a solution generated by the CMI always fulfills Assumption 3.2.

Lemma 3.7

For a solution $\vec{X}_{\mathcal{I}^+}^{CMI}$ to the MRTA task insertion reoptimization problem (Problem 3.3) generated by the application of the CMI (see Algorithm 1), it always holds that

$$\Delta^+ \leq J_{\mathcal{I}^+}^*. \tag{3.20}$$

Proof:
Consider an unknown globally optimal solution $\vec{X}_{\mathcal{I}^+}^*$ of the modified problem instance $\mathcal{I}^+$. Within the solution $\vec{X}_{\mathcal{I}^+}^*$, the task n_{N+1} is assumed to be allocated to an agent $m^* \in \mathcal{K}$. Then the value of the ojective function $J_{\mathcal{I}^+}^*$ of the solution $\vec{X}_{\mathcal{I}^+}^*$ can be written as

$$J_{\mathcal{I}^+}^* = \phi_{\mathcal{K} \setminus m^*} + \phi_{m^*} \tag{3.21}$$

where ϕ_{m^*} denotes the objective function value associated with the route of agent m^* (i. e. the weighted sum of task execution, transitioning and waiting times of agent m^*, see objective function (3.1)) and $\phi_{\mathcal{K} \setminus m^*}$ denotes the sum of the objective function values associated with the routes of all other agents $m \in \mathcal{K} \setminus m^*$. It is known that the route of agent m^* starts and ends at the depot o_{m^*} and contains at least task n_{N+1}. Consequently, its objective function value at least contains the task execution time for executing task n_{N+1} plus the times needed for the transition from its depot o_{m^*} to task n_{N+1} and back. Using the triangle inequality and the symmetry of distances it follows for the objective function value associated with the unknown route of agent m^*

that

$$\phi_{m^*} \geq \frac{\delta(o_{m^*}, n_{N+1}) + \delta(n_{N+1}, o_{m^*})}{v_{m^*}} + \gamma \frac{\tau_{N+1}}{a^{m^*}_{N+1}} = \frac{2\delta(n_{N+1}, o_{m^*})}{v_{m^*}} + \gamma \frac{\tau_{N+1}}{a^{m^*}_{N+1}}. \quad (3.22)$$

In addition to the underapproximation (3.22) of the objective function value associated with the unknown route of agent m^*, an overapproximation of the optimal maximum insertion cost is derived. To this end, (3.19) from Lemma 3.6 as well as the definition of the optimal maximum insertion cost which is given by inserting (3.13) into (3.15) are used. The overapproximation of the optimal maximum insertion cost follows, which holds for inserting task n_{N+1} on any edge (i, j) of the route of any agent $m \in \mathcal{K}$ that is chosen in the routing $X^*_{\mathcal{I}}$ of the initial solution $\vec{X}^*_{\mathcal{I}}$

$$\Delta^*_{\max} \overset{(3.19)}{\leq} 2\frac{\delta(n_{N+1}, j)}{v_m} + \gamma \frac{\tau_{N+1}}{a^m_{N+1}} + \epsilon\left(\beta_j - 1\right)\left(2\frac{\delta(n_{N+1}, j)}{v_m} + \frac{\tau_{N+1}}{a^m_{N+1}}\right)$$
$$\forall \left\{(j, m) \,\middle|\, \exists x^m_{i,j} \in X^*_{\mathcal{I}} : x^m_{i,j} = 1\right\}. \quad (3.23)$$

Since (3.23) holds for inserting task n_{N+1} on any edge (i, j) of the route of any agent $m \in \mathcal{K}$ within the initial routing $X^*_{\mathcal{I}}$, it is also fulfilled for inserting task n_{N+1} on the last edge of the route of agent m^*, i. e. for $j = o_{m^*}$ and $m = m^*$. Inserting task n_{N+1} on the last edge of the route of agent m^* cannot introduce any additional waiting times since there is no subsequent task in agent m^*'s route and thus also no task with potential precedence or synchronization constraints can be temporally shifted, which is why $\beta_{o_{m^*}} = 1$ holds. Inserting these values in (3.23) yields

$$\Delta^*_{\max} \leq 2\frac{\delta(n_{N+1}, o_{m^*})}{v_{m^*}} + \gamma \frac{\tau_{N+1}}{a^{m^*}_{N+1}}. \quad (3.24)$$

From (3.22) and (3.24), if follows that $\Delta^*_{\max} \leq \phi_{m^*}$. As given by (3.17), the optimal maximum insertion costs $\Delta^*_{\max}$ are always an upper bound for the real increment in the objective function Δ^+. Furthermore, $\phi_{m^*} \leq J^*_{\mathcal{I}^+}$ follows from (3.21). Thus,

$$\Delta^+ \leq \Delta^*_{\max} \leq \phi_{m^*} \leq J^*_{\mathcal{I}^+} \quad (3.25)$$

results. □

Using Lemma 3.7, the following guarantee can be given for the CMI:

Theorem 3.2

Solving an MRTA task insertion reoptimization problem (Problem 3.3), for which Assumption 3.1 holds, by the application of the CMI (see Algorithm 1) leads to an approximation ratio

$$\alpha = \frac{J^{CMI}_{\mathcal{I}^+}}{J^*_{\mathcal{I}^+}} \leq 2. \quad (3.26)$$

Proof:
According to Lemma 3.7, Assumption 3.2 is always fulfilled for a solution to the MRTA task insertion reoptimization problem generated by the CMI. Thus, by applying Theorem 3.1, the assertion follows. □

Furthermore, the upper bound on the approximation ratio of $\alpha \leq 2$ is the smallest bound possible as given by the following proposition.

Proposition 3.1
For the application of the CMI as defined in Algorithm 1 to the task insertion reoptimization problem (Problem 3.3), the bound on the approximation ratio of $\alpha \leq 2$ is tight, i. e. no lower upper bound on the approximation ratio exists.

Proof:
To prove that $\alpha \leq 2$ is the smallest upper bound possible on the approximation ratio resulting from the application of the CMI to MRTA task insertion reoptimization problems, it is shown that an MRTA task insertion reoptimization problem instance with an approximation ratio converging towards $\alpha = 2$ resulting from the application of the CMI exists. It is depicted in Figure 3.5. An initial MRTA problem instance $\mathcal{I}$ is given which consists of two agents having a transition velocity of $v_1 = 1$ and $v_2 = 1/2$. The agents' depots are located in a distance of $\delta(o_1, n_1) = 1$ and $\delta(o_2, n_1) = 1/2 - \eta$, $0 \leq \eta \leq 1/2$ from the task n_1, for which both have a capability of $a_1^1 = a_1^2 = 1$.

Both possible routings to this instance, i. e. either allocating task n_1 to agent k_1 or to agent k_2, have the same task execution time. For $\eta > 0$, the transition time required by agent k_2 is smaller than the one required by agent k_1. Since furthermore, both possible solutions corresponding to these routings have no waiting times, the globally optimal solution is given by routing agent k_2 to task n_1 which results in an objective function value of $J_{\mathcal{I}}^* = 2 - 4\eta$. In the modified instance $\mathcal{I}^+$, task n_2 is added at the same position as task n_1. Since only agent k_1 is capable of executing task n_2, i. e. $a_2^1 = 1$ and $a_2^2 = 0$, task n_2 is assigned to agent k_1 in the solution generated by the CMI which corresponds to an objective function value of $J_{\mathcal{I}^+}^{\text{CMI}} = 4 - 4\eta$. The optimal solution however would assign both tasks to agent k_1, which would lead to an optimal objective function value of $J_{\mathcal{I}^+}^* = 2$, i. e. the resulting approximation ratio equals $\alpha = 2 - 2\eta$ which converges towards $\alpha \to 2$ for $\eta \to 0$. Given that a problem instance exists, for which the approximation ratio converges towards 2 from below, no smaller upper bound can be given for the CMI. □

The approximation ratio guaranteed for the CMI so far holds for MRTA task insertion reoptimization problems according to Problem 3.3 in which the initial problem instance $\mathcal{I}$ can have all features as introduced in Section 3.1.1. Thus, agents can have different

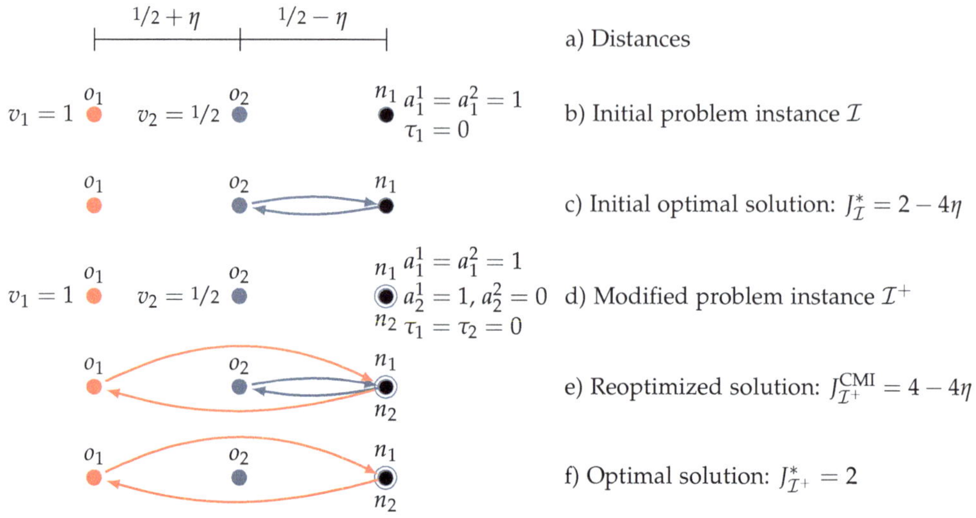

Figure 3.5: Example of an MRTA task insertion reoptimization problem instance with an approximation ratio of $\alpha \to 2$ for $\eta \to 0$ for the application of the CMI. The figure is based on the author's publication [BKH+24].

capabilities and transition velocities and precedence and synchronization constraints can be considered. If temporally unconstrained problem instances or even homogeneous groups of agents are considered, even smaller upper bounds on the approximation ratio can be guaranteed. Before starting this analysis, the considered problem configurations are introduced in the following.

Considered problem configurations

To reveal the difference in guaranteed approximation ratios for the MRTA task insertion reoptimization problem (Problem 3.3) dependent on the properties of the MRTA problem instance $\mathcal{I}^+$, three problem configurations are introduced. They differ w. r. t. the features of the corresponding MRTA problems. The problem configuration introduced in Section 3.1.1 that has been considered so far is denoted as $P_{\mathcal{P},\mathcal{S},\mathrm{het}}$. It allows for the consideration of fully heterogeneous teams of agents as well as both types of temporal constraints, i. e. precedence and synchronization constraints, and therefore imposes no additional restrictions on the MRTA task insertion reoptimization problem defined in Problem 3.3. In contrast to this, the configurations P_{het} and P_{hom} assume all tasks $i \in \mathcal{N}$ to be temporally unconstrained. Thus, all precedence and synchronization constraints are assumed to be zero, i. e. $p_{i,j} = 0$, $\forall p_{i,j} \in \mathcal{P}, p_{i,j} \in \mathcal{P}^+$ and $s_{i,j} = 0$, $\forall s_{i,j} \in \mathcal{S}, s_{i,j} \in \mathcal{S}^+$. The problem configuration with heterogeneous teams of agents P_{het} allows for agents that differ both in transition velocities and in task execution capabilities. In homogeneous problem instances of configuration P_{hom} however, all agents are assumed to have transition velocities of one and task execution capabilities equal to one for all tasks. An overview of the problem configurations is depicted in Table 3.2.

Remark. *Within the problem configuration P_{hom}, setting the homogeneous velocities and capabilities each to the value of one can be done without loss of generality, since any problem instance with homogeneous transition velocities and task execution capabilities can be transformed to this form by normalization.*

CMI for temporally unconstrained MRTA task insertion reoptimization problems

In the problem configurations P_{het} and P_{hom} neither precedence nor synchronization constraints are considered. Consequently, no waiting times can occur within a solution. The maximum insertion costs $\Delta_{\text{max}}^{i,j,m}$ calculated within the CMI (see Algorithm 1) therefore equal the actual increase in the objective function $\Delta^{i,j,m}$ associated with the insertion of the new task n_{N+1} on edge (i,j) of the route of agent m. In other words, without any precedence and synchronization constraints, the maximum number of agents' routes affected by a temporal shift of any task $i \in \mathcal{N}$ is always equal to one, i. e. $\beta_i = 1, \forall i \in \mathcal{N}$ (see (3.16)). From (3.13) it therefore follows for the CMI applied to problems of the configurations P_{het} and P_{hom}

$$P_{\text{het}}, P_{\text{hom}} : \Delta_{\text{max}}^{i,j,m} = \Delta^{i,j,m} = \begin{cases} \Delta_{\text{trans}}^{i,j,m} + \gamma d_{N+1}^{m} = \Delta_{\text{trans}}^{i,j,m} + \gamma \frac{\tau_{N+1}}{a_{N+1}^{m}} & \text{if } a_{N+1}^{m} > 0 \\ \infty & \text{if } a_{N+1}^{m} = 0 \end{cases}$$

$$\forall (i,j,m) : x_{i,j}^{m} \in X_{\mathcal{I}}^{*} \wedge x_{i,j}^{m} = 1 \quad (3.27)$$

Since no overestimation of the maximum insertion costs takes place for problems of configurations P_{het} and P_{hom}, the optimal maximum insertion costs Δ_{max}^{*} determined by the CMI are equivalent to the actual increase in the objective function Δ^{+} of the

Table 3.2: Problem configurations considered for reoptimization.

Problem configuration	Assumptions on velocities $\mathcal{V}$	task capabilities $\mathcal{A}$	precedence constraints $\mathcal{P}$	synchronization constraints $\mathcal{S}$
$P_{\mathcal{P},\mathcal{S},\text{het}}$	—	—	—	—
P_{het}	—	—	$p_{i,j} = 0$, $\forall p_{i,j} \in \mathcal{P} \cup \mathcal{P}^{+}$	$s_{i,j} = 0$, $\forall s_{i,j} \in \mathcal{S} \cup \mathcal{S}^{+}$
P_{hom}	$v_m = 1$, $\forall v_m \in \mathcal{V}$	$a_i^m = 1$, $\forall a_i^m \in \mathcal{A} \cup \mathcal{A}^{+}$	$p_{i,j} = 0$, $\forall p_{i,j} \in \mathcal{P} \cup \mathcal{P}^{+}$	$s_{i,j} = 0$, $\forall s_{i,j} \in \mathcal{S} \cup \mathcal{S}^{+}$

solution generated by the CMI compared to the objective function value of the initial solution, i. e.

$$P_{\text{het}}, P_{\text{hom}} : \Delta^*_{\max} = \min_{\left\{(i,j,m) \middle| x^m_{i,j} \in X^*_{\mathcal{I}} \wedge x^m_{i,j}=1\right\}} \Delta^{i,j,m} =: \Delta^* = \Delta^+. \tag{3.28}$$

Remark. *For an unambiguous notation, the following distinction is made: Δ^+ denotes the actual increase in the objective function, independent of the configuration of the problem instance and of the reoptimization approach applied. For the application of the CMI, the potential overestimate of the increase in the objective function of the CMI, when applied to a general problem instance of configuration $P_{\mathcal{P},\mathcal{S},het}$ that might contain precedence and synchronization constraints, is denoted by Δ^*_{max} (see* (3.13)*). If the CMI is known to be applied to a problem instance of configurations P_{het} or P_{hom}, for which* (3.27) *and* (3.28) *hold, the increase in the objective function resulting from the application of the CMI is denoted as Δ^*, which is known to equal the actual cost increment Δ^+.*

Remark. *Since problem instances of configurations P_{het} and P_{hom} are special cases of the general Problem 3.3 of configuration $P_{\mathcal{P},\mathcal{S},het}$, the solutions determined by the CMI for problems of configurations P_{het} and P_{hom} are also guaranteed to be feasible according to Lemma 3.5.*

In the following, the approximation ratios resulting from applying the CMI to temporally unconstrained task insertion reoptimization problems of configurations P_{het} and P_{hom} are analyzed.

Analysis of temporally unconstrained MRTA task insertion reoptimization problems

To analyze performance guarantees in the form of upper bounds on the approximation ratio for the CMI when applied to temporally unconstrained but heterogeneous problem instances according to configuration P_{het}, the maximum and minimum velocity of the agents in the considered problem instance are of interest. They are denoted by $v^{\max} := \max_{\{v \in \mathcal{V}\}}(v)$ and $v^{\min} := \min_{\{v \in \mathcal{V}\}}(v)$, respectively. Furthermore, the maximum and the minimum of the agents' real valued capabilities to perform the task n_{N+1} are denoted by $a^{\max}_{N+1} := \max_{\{m \in \mathcal{K}\}}\left(a^m_{N+1}\right)$ and $a^{\min}_{N+1} := \min_{\{m \in \mathcal{K}\}}\left(a^m_{N+1}\right)$. Using this notation, the following theorem for upper bounds on the approximation ratio holds.

Theorem 3.3

Solving an MRTA task insertion reoptimization problem (Problem 3.3) of problem configuaration P_{het} (see Table 3.2) by the application of the CMI (see Algorithm 1), leads to an approximation ratio

$$P_{het}: \quad \alpha = \frac{J_{\mathcal{I}^+}^{CMI}}{J_{\mathcal{I}^+}^{*}} \leq \min\left\{ \frac{3v^{\max}a_{N+1}^{\max}}{2v^{\min}a_{N+1}^{\min}}, 2 \right\}. \tag{3.29}$$

Proof:

The set of problem instances of configuration P_{het} is a subset of the set of problem instances of the general configuration $P_{\mathcal{P},\mathcal{S},\text{het}}$. Since neither precedence nor synchronization are contained within problem instances of configuration P_{het}, Assumption 3.1 is always fulfilled according to Lemma 3.4. Consequently, Theorem 3.2 holds for all heterogeneous task insertion reoptimization problems without precedence and synchronization constraints of configuration P_{het}. Furthermore, in the following

$$\alpha \leq \frac{3v^{\max}a_{N+1}^{\max}}{2v^{\min}a_{N+1}^{\min}}$$

is proven to hold for problems of configuration P_{het}.

Without loss of generality, the agents' velocities and their capabilities for task n_{N+1} as well as all distances and basic task durations can be normalized, i. e.

$$\overline{v}_m := \frac{v_m}{v^{\max}} \quad \Rightarrow \quad \overline{v}^{\max} = 1, \quad \overline{v}^{\min} = \frac{v^{\min}}{v^{\max}} \tag{3.30}$$

$$\overline{\delta}(i,j) := \frac{\delta(i,j)}{v^{\max}} \tag{3.31}$$

$$\overline{a}_{N+1}^{m} := \frac{a_{N+1}^{m}}{a_{N+1}^{\max}} \quad \Rightarrow \quad \overline{a}_{N+1}^{\max} = 1, \quad \overline{a}_{N+1}^{\min} = \frac{a_{N+1}^{\min}}{a_{N+1}^{\max}} \tag{3.32}$$

$$\overline{\tau}_{N+1} := \frac{\tau_{N+1}}{a_{N+1}^{\max}} \tag{3.33}$$

Using the CMI insertion costs Δ^* that apply for problems of configuration P_{het} as given by (3.27) and (3.28) as well as (3.19) from Lemma 3.6, it follows

$$\begin{aligned}
\Delta^* &\leq \frac{\overline{\delta}(i,n_{N+1}) + \overline{\delta}(n_{N+1},j) - \overline{\delta}(i,j)}{\overline{v}_m} + \gamma\frac{\overline{\tau}_{N+1}}{\overline{a}_{N+1}^{m}} \\
&\qquad\qquad\qquad\qquad \forall(i,j,m) : x_{i,j}^{m} \in X_{\mathcal{I}}^{*} \wedge x_{i,j}^{m} = 1 \\
&\overset{(3.19)}{\leq} \frac{2\overline{\delta}(i,n_{N+1})}{\overline{v}_m} + \gamma\frac{\overline{\tau}_{N+1}}{\overline{a}_{N+1}^{m}} \quad \forall(i,m) : \left(\exists j \in V : x_{i,j}^{m} \in X_{\mathcal{I}}^{*} \wedge x_{i,j}^{m} = 1\right) \\
&\leq \frac{2\overline{\delta}(i,n_{N+1})}{\overline{v}^{\min}} + \gamma\frac{\overline{\tau}_{N+1}}{\overline{a}_{N+1}^{\min}} \quad \forall i \in V.
\end{aligned} \tag{3.34}$$

Let i^* be the vertex precedent to task n_{N+1} within an unknown globally optimal solution $\vec{X}^*_{\mathcal{I}^+}$ of the modified instance $\mathcal{I}^+$ and j^* the vertex following task n_{N+1}, respecitvely. The vertices i^*, j^* and n_{N+1} are assumed to be allocated to agent m^* in the routing $X^*_{\mathcal{I}^+}$ of solution $\vec{X}^*_{\mathcal{I}^+}$. Evaluating (3.34) for vertices i^* and j^* and summing up both inequalities yields

$$\begin{aligned} 2\Delta^* &\leq 2\frac{\overline{\delta}(i^*, n_{N+1})}{\overline{v}^{\min}} + 2\frac{\overline{\delta}(j^*, n_{N+1})}{\overline{v}^{\min}} + 2\gamma\frac{\overline{\tau}_{N+1}}{\overline{a}^{\min}_{N+1}} \\ \Leftrightarrow \quad \Delta^* &\leq \frac{\overline{\delta}(i^*, n_{N+1})}{\overline{v}^{\min}} + \frac{\overline{\delta}(n_{N+1}, j^*)}{\overline{v}^{\min}} + \gamma\frac{\overline{\tau}_{N+1}}{\overline{a}^{\min}_{N+1}} \end{aligned} \tag{3.35}$$

Within the optimal solution $\vec{X}^*_{\mathcal{I}^+}$, let $\varphi_{m^*}(i^*, j^*)$ denote the value of the objective function associated with the route of agent m^* without the part between the vertices i^* and j^*, i. e. without the execution of task n_{N+1} and without the transition from i^* to n_{N+1} and from n_{N+1} to j^*. Then the objective function (3.1) of the unknown optimal solution $J^*_{\mathcal{I}^+}$ can be written as

$$J^*_{\mathcal{I}^+} = \phi_{\mathcal{K}\backslash m^*} + \varphi_{m^*}(i^*, j^*) + \frac{\overline{\delta}(i^*, n_{N+1})}{\overline{v}_{m^*}} + \frac{\overline{\delta}(n_{N+1}, j^*)}{\overline{v}_{m^*}} + \gamma\frac{\overline{\tau}_{N+1}}{\overline{a}^{m^*}_{N+1}}. \tag{3.36}$$

Using the knowledge of agent m^* being allocated at least to vertices i^*, j^* and to task n_{N+1} within the globally optimal solution $\vec{X}^*_{\mathcal{I}^+}$, it follows that $\varphi_{m^*}(i^*, j^*)$ contains at least the tranistioning of agent m^* from its depot o_{m^*} to i^* and from j^* to o_{m^*}. Together with the triangle inequality,

$$\begin{aligned} \varphi_{m^*}(i^*, j^*) &\geq \frac{\overline{\delta}(o_{m^*}, i^*) + \overline{\delta}(j^*, o_{m^*})}{\overline{v}_{m^*}} \\ &\geq \frac{\overline{\delta}(i^*, j^*)}{\overline{v}_{m^*}} \end{aligned} \tag{3.37}$$

follows. Furthermore, just skipping task n_{N+1} within the routing $X^*_{\mathcal{I}^+}$ of the unknown solution $\vec{X}^*_{\mathcal{I}^+}$ to the modified problem instance $\mathcal{I}^+$ yields a routing that together with corresponding optimized timing information (see (3.5)) yields a solution to the initial problem instance $\mathcal{I}$, that by definition cannot have a smaller objective function value than the globally optimal one. Thus, the following overestimation of the objective function value $J^*_{\mathcal{I}}$ of the globally optimal initial solution holds:

$$\begin{aligned} J^*_{\mathcal{I}} &\leq \phi_{\mathcal{K}\backslash m^*} + \varphi_{m^*}(i^*, j^*) + \frac{\overline{\delta}(i^*, j^*)}{\overline{v}_{m^*}} \overset{(3.37)}{\leq} \phi_{\mathcal{K}\backslash m^*} + 2\varphi_{m^*}(i^*, j^*) \\ &\leq 2\left(\phi_{\mathcal{K}\backslash m^*} + \varphi_{m^*}(i^*, j^*)\right) \end{aligned} \tag{3.38}$$

Equation (3.36) together with (3.38) yields

$$J^*_{\mathcal{I}^+} \geq \frac{J^*_{\mathcal{I}}}{2} + \frac{\overline{\delta}(i^*, n_{N+1})}{\overline{v}_{m^*}} + \frac{\overline{\delta}(n_{N+1}, j^*)}{\overline{v}_{m^*}} + \gamma\frac{\overline{\tau}_{N+1}}{\overline{a}^{m^*}_{N+1}}. \tag{3.39}$$

Two cases are considered to finish the proof that $\alpha \leq \left(3v^{\max}a_{N+1}^{\max}\right)/\left(2v^{\min}a_{N+1}^{\min}\right)$ holds:

Case 1: $\overline{\delta}(i^*, n_{N+1}) + \overline{\delta}(n_{N+1}, j^*) + \gamma\overline{\tau}_{N+1} \geq \frac{J_{\mathcal{I}}^*}{2}$:

$$\begin{aligned}
\alpha = \frac{J_{\mathcal{I}}^* + \Delta^*}{J_{\mathcal{I}^+}^*} &\overset{(3.35),(3.39)}{\leq} \frac{J_{\mathcal{I}}^* + \frac{1}{\overline{v}^{\min}}\left(\overline{\delta}(i^*, n_{N+1}) + \overline{\delta}(n_{N+1}, j^*)\right) + \gamma\frac{1}{\overline{a}_{N+1}^{\min}}\overline{\tau}_{N+1}}{\frac{J_{\mathcal{I}}^*}{2} + \frac{1}{v_{m^*}}\left(\overline{\delta}(i^*, n_{N+1}) + \overline{\delta}(n_{N+1}, j^*)\right) + \gamma\frac{1}{\overline{a}_{N+1}^{m^*}}\overline{\tau}_{N+1}} \\
&\overset{\left(\substack{\overline{v}^{\max}=1 \\ \overline{a}_{N+1}^{\max}=1}\right)}{\leq} \frac{J_{\mathcal{I}}^* + \frac{1}{\overline{v}^{\min}}\left(\overline{\delta}(i^*, n_{N+1}) + \overline{\delta}(n_{N+1}, j^*)\right) + \gamma\frac{1}{\overline{a}_{N+1}^{\min}}\overline{\tau}_{N+1}}{\frac{J_{\mathcal{I}}^*}{2} + \left(\overline{\delta}(i^*, n_{N+1}) + \overline{\delta}(n_{N+1}, j^*)\right) + \gamma\overline{\tau}_{N+1}} \\
&\leq \frac{1}{\overline{v}^{\min}\,\overline{a}_{N+1}^{\min}} \frac{J_{\mathcal{I}}^* + \overline{\delta}(i^*, n_{N+1}) + \overline{\delta}(n_{N+1}, j^*) + \gamma\overline{\tau}_{N+1}}{\frac{J_{\mathcal{I}}^*}{2} + \overline{\delta}(i^*, n_{N+1}) + \overline{\delta}(n_{N+1}, j^*) + \gamma\overline{\tau}_{N+1}} \\
&\overset{(\text{Case 1})}{\leq} \frac{3}{2}\frac{1}{\overline{v}^{\min}\,\overline{a}_{N+1}^{\min}} = \frac{3}{2}\frac{v^{\max}a_{N+1}^{\max}}{v^{\min}a_{N+1}^{\min}}
\end{aligned}$$

Case 2: $\overline{\delta}(i^*, n_{N+1}) + \overline{\delta}(n_{N+1}, j^*) + \gamma\overline{\tau}_{N+1} < \frac{J_{\mathcal{I}}^*}{2}$:

$$\begin{aligned}
\alpha = \frac{J_{\mathcal{I}}^* + \Delta^*}{J_{\mathcal{I}^+}^*} &\overset{(3.6),(3.35)}{\leq} \frac{J_{\mathcal{I}}^* + \frac{1}{\overline{v}^{\min}}\left(\overline{\delta}(i^*, n_{N+1}) + \overline{\delta}(n_{N+1}, j^*)\right) + \gamma\frac{1}{\overline{a}_{N+1}^{\min}}\overline{\tau}_{N+1}}{J_{\mathcal{I}}^*} \\
&\leq \frac{1}{\overline{v}^{\min}\overline{a}_{N+1}^{\min}} \frac{J_{\mathcal{I}}^* + \left(\overline{\delta}(i^*, n_{N+1}) + \overline{\delta}(n_{N+1}, j^*)\right) + \gamma\overline{\tau}_{N+1}}{J_{\mathcal{I}}^*} \\
&\overset{(\text{Case 2})}{<} \frac{3}{2}\frac{1}{\overline{v}^{\min}\,\overline{a}_{N+1}^{\min}} = \frac{3}{2}\frac{v^{\max}a_{N+1}^{\max}}{v^{\min}a_{N+1}^{\min}}
\end{aligned}$$

Since both, $\alpha \leq \left(3v^{\max}a_{N+1}^{\max}\right)/\left(2v^{\min}a_{N+1}^{\min}\right)$ as well as $\alpha \leq 2$ must hold, (3.29) follows. □

> **Proposition 3.2**
>
> *For the application of the CMI as defined in Algorithm 1 to the task insertion reoptimization problem (Problem 3.3) of problem configuaration P_{het} (see Table 3.2), the bound on the approximation ratio of $\alpha \leq 2$ is tight, i. e. no lower upper bound on the approximation ratio exists.*

Proof:
The task insertion reoptimization problem instance depicted in Figure 3.5 is of configuration P_{het}, since no precedence or synchronization constraints are considered. Consequently, following the proof of Proposition 3.1, $\alpha \leq 2$ is a tight bound for the approximation ratio of adding a task to a problem instance of configuration P_{het}. □

Using the result obtained for MRTA task insertion reoptimization problem instances of configuration level P_{het}, a smaller upper bound on the approximation ratio resulting from the application of the CMI to homogeneous problem instances of configuration P_{hom} is given in the following.

Corollary 3.2

Solving an MRTA task insertion reoptimization problem (Problem 3.3) of problem configuaration P_{hom} (see Table 3.2) by the application of the CMI as defined in Algorithm 1, leads to an approximation ratio

$$P_{hom}: \quad \alpha = \frac{J_{\mathcal{I}^+}^{CMI}}{J_{\mathcal{I}^+}^{*}} \leq \frac{3}{2}. \tag{3.40}$$

Proof:

Since problem instances of configuration P_{hom} are a subset of the problem instances of configuration P_{het} (see Table 3.2), Theorem 3.3 holds for homogeneous MRTA task insertion reoptimization instances of problem configuration P_{hom}. By definition of configuration P_{hom}, the velocities and task execution capabilities of all agents are equal, i. e. $v^{\min} = v^{\max}$ and $a_{N+1}^{\min} = a_{N+1}^{\max}$. By inserting this into (3.29) of Theorem 3.3, the assertion follows. □

For homogeneous MRTA task insertion reoptimization problems, the upper bound on the approximation ratio proven for the CMI is furthermore tight.

Proposition 3.3

For the application of the CMI as defined in Algorithm 1 to the task insertion reoptimization problem (Problem 3.3) of problem configuaration P_{hom} (see Table 3.2), the bound on the approximation ratio of $\alpha \leq \frac{3}{2}$ is tight, i. e. no lower upper bound on the approximation ratio exists.

Proof:

To prove that $\alpha \leq 3/2$ is the smallest upper bound possible on the approximation ratio resulting from the application of the CMI to homogeneous MRTA task insertion reoptimization problems of configuration P_{hom}, a homogeneous MRTA task insertion reoptimization problem instance with an approximation ratio converging towards $\alpha \to 3/2$ for $\eta \to 0$, $\eta \geq 0$, resulting from the application of the CMI is depicted in Figure 3.6. Since the depicted problem instance is of configuration P_{hom}, the velocities and capabilities are given by $v_1 = v_2 = 1$ and $a_1^1 = a_1^2 = a_2^1 = a_2^2 = 1$. An initial MRTA problem instance $\mathcal{I}$ is depicted which consists of two agents being located with their depots in a distance of $\delta(o_1, n_1) = 2$ and $\delta(o_2, n_1) = 2 - \eta$ for $\eta \geq 0$ of the task n_1. Due to the slightly shorter distance from depot o_2, task n_1 is allocated to agent k_2 in the initial solution $\vec{X}^*_{\mathcal{I}}$, which has an objective function value of $J^*_{\mathcal{I}} = 4 - 2\eta$. In the modified instance $\mathcal{I}^+$, task n_2 is added between task n_1 and depot o_1. As depicted in Figures 3.6 e) and in Figures 3.6 f), two CMI solutions of the same objective function value of $J^{\text{CMI}}_{\mathcal{I}^+} = 6 - 2\eta$ exist. The optimal solution however would be to assign both tasks to agent k_1, which would lead to an optimal objective function value of $J^*_{\mathcal{I}^+} = 4$. Thus, for $\eta \rightarrow 0$, the approximation ratio converges towards $3/2$, i. e. $\alpha \rightarrow 3/2$ for $\eta \rightarrow 0$ and consequently no upper bound on the approximation ratio smaller than $3/2$ exists for homogeneous MRTA task insertion reoptimization problem instances of configuration P_{hom}. □

Summary of the performance guarantees for the CMI

The CMI (see Algorithm 1), introduced to solve the MRTA task insertion reoptimization problem (Problem 3.3), overapproximates the additional waiting times when inserting a task n_{N+1} to an initial solution $\vec{X}^*_{\mathcal{I}}$. Any solution found by the CMI to an MRTA task insertion reoptimization problem is guaranteed to be feasible as proven by Theorem 3.5.

Furthermore, despite the overapproximation of the additional waiting times resulting from inserting an additional task n_{N+1} to an initial solution $\vec{X}^*_{\mathcal{I}}$, tight bounds on the approximation ratio can be given. Thus, the CMI is the first task insertion reoptimization heuristic for heterogeneous, time-extended MRTA problems with precedence and synchronization constraints for which performance guarantees in the form of upper bounds on the approximation ratio can be given. Depending on the problem configuration of the problem instances considered, these bounds are even smaller than the general upper bound of $\alpha \leq 2$ which is proven in Section 3.2.2. These results are summarized in Table 3.3. For the full problem configuration $P_{\mathcal{P},\mathcal{S},\text{het}}$ that allows for precedence and synchronization constraints as well as for heterogeneous agents, a tight upper bound of $\alpha \leq 2$ is guaranteed by Theorem 3.2 and Proposition 3.1. If no precedence and synchronization constraints are considered as in problem configuration P_{het}, according to Theorem 3.3 the guaranteed upper bound on the approximation ratio varies between $\alpha \leq 3/2$ and $\alpha \leq 2$ and depends on the velocity differences of the agents as well as on the differences in their capabilities to perform the inserted task n_{N+1}. For problem instances with homogeneous groups of agents and without precedence and synchronization constraints, a tight upper bound of $\alpha \leq 3/2$ is guaranteed by Theorem 3.2 and Proposition 3.3.

In the following section, the reoptimization problem of deleting a task from an MRTA problem instance is defined and analyzed.

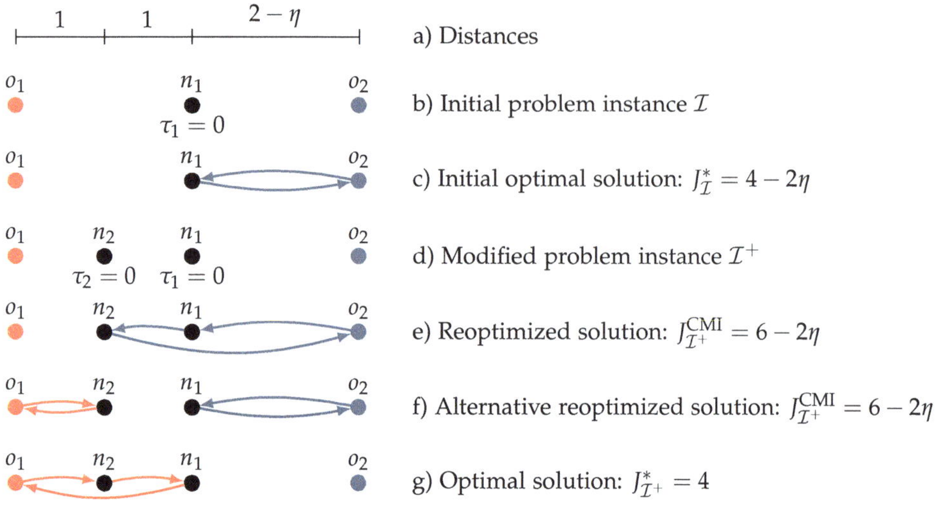

Figure 3.6: Example of a homogeneous MRTA task insertion reoptimization problem instance of configuration P_{hom} with an approximation ratio of $\alpha \to 3/2$ for $\eta \to 0$ for the application of the CMI. Since the depicted problem instance is of configuration P_{hom}, the velocities and capabilities are given by $v_1 = v_2 = 1$ and $a_1^1 = a_1^2 = a_2^1 = a_2^2 = 1$, respectively. The figure is based on the author's publication [BKH+24].

Table 3.3: Guaranteed upper bounds on the approximation ratios for the application of the CMI to MRTA task insertion reoptimization problems.

Problem configuration	Approximation ratio α	Tight bound
$P_{\mathcal{P},\mathcal{S},\text{het}}$	≤ 2	yes
P_{het}	$\leq \min\left\{\frac{3v^{\max}a_{N+1}^{\max}}{2v^{\min}a_{N+1}^{\min}}, 2\right\}$	tight for $\alpha \leq 2$
P_{hom}	$\leq \frac{3}{2}$	yes

3.3 Task Deletion

In this section, the task deletion reoptimization problem corresponding to the time-extended MRTA optimization problem defined in Problem 3.1 is both introduced and analyzed. A performance guarantee in the form of an upper bound of 2 on the approximation ratio is given which holds for any reoptimization approach fulfilling two assumptions on the resulting difference in objective function value between the initial and the reoptimized solution. Subsequently, the *task deletion heuristic (TDH)* is introduced, and it is shown, that the TDH fulfills the respective assumptions for the vast majority of problem instances. Furthermore, a tight upper bound of $\alpha \leq 3/2$ is proven for temporally unconstrained, homogeneous MRTA task deletion reoptimization problems in which the agents originate from one common depot. The results presented in the following are based on the author's journal publication [BKH+24].

3.3.1 Task Deletion Reoptimization Problem

The MRTA task deletion reoptimization problem is, given a globally optimal solution $\vec{X}^*_{\mathcal{I}}$ to an initial problem instance $\mathcal{I}$, how to use this solution to solve a related problem instance $\mathcal{I}^-$ which differs from the initial instance $\mathcal{I}$ by one task n_q that has been deleted from the initial problem instance. A comprehensive formal problem definition is given in the following.

Problem 3.4 (MRTA task deletion reoptimization problem)
The MRTA task deletion reoptimization problem is an MRTA reoptimization problem according to Problem 3.2, where the modified problem instance $\mathcal{I}^-$ is given by $\mathcal{I}^- = \{\mathcal{N}^-, \mathcal{K}, \mathcal{O}, \mathcal{V}, \mathcal{A}^-, \mathcal{T}^-, \mathcal{D}^-, \mathcal{P}^-, \mathcal{S}^-\}$ with the modified sets

- $\mathcal{N}^- = \mathcal{N} \setminus \{n_q\}$,
- $\mathcal{A}^- = \mathcal{A} \setminus \left\{a_q^m \,|\forall m \in \mathcal{K}\right\}$,
- $\mathcal{T}^- = \mathcal{T} \setminus \{\tau_q\}$,
- $\mathcal{D}^- = \mathcal{D} \setminus \{\delta(i, n_q), \delta(n_q, i) \,|\forall i \in V\}$,
- $\mathcal{P}^- = \mathcal{P} \setminus \{p_{i,q}, p_{q,i} \,|\forall\, i \in \mathcal{N}\}$ *and*
- $\mathcal{S}^- = \mathcal{S} \setminus \{s_{i,q}, s_{q,i} \,|\forall\, i \in \mathcal{N}\}$.

Thus, the modified problem instance $\mathcal{I}^-$ misses one task n_q and all corresponding precedence and synchronization constraints compared to the initial problem instance $\mathcal{I}$.

A reoptimization approach independent guarantee on the solution quality for the MRTA task deletion reoptimization problem (Problem 3.4) is given in the following section.

3.3.2 Reoptimization Approach Independent Guarantee on the Approximation Ratio

In this section, it is shown that any solution approach that solves the MRTA task deletion reoptimization problem (Problem 3.4) and that fulfills two assumptions on the resulting difference in objective function value between the initial and the reoptimized solution, cannot yield a reoptimization ratio greater than 2.

To analyze the approximation ratio, the following considerations are relevant: When solving a task deletion reoptimization problem, an optimal solution $\vec{X}^*_{\mathcal{I}}$ of the initial problem instance $\mathcal{I}$ is used to derive a heuristic solution to the modified problem instance $\mathcal{I}^-$ which differs from the initial instance by the task set not containing task n_q as defined in Problem 3.4. The objective function value of the generated reoptimized solution of the modified problem instance $J_{\mathcal{I}^-}^{\text{reo}}$ and the value of the objective function of the initial solution $J_{\mathcal{I}}^*$ differ by a cost difference Δ^-, i. e.

$$J_{\mathcal{I}^-}^{\text{reo}} = J_{\mathcal{I}}^* - \Delta^-. \tag{3.41}$$

Depending on the specific reoptimization method used to solve the task deletion reoptimization problem (Problem 3.4), the difference in objective function value Δ^- resulting from the application of the respective method to an instance of the task deletion reoptimization problem may differ. The following assumption is made on Δ^-:

Assumption 3.3. *Let the task deletion reoptimization approach, the initial problem instance $\mathcal{I}$ and the modified problem instance $\mathcal{I}^-$ be defined such that the reduction Δ^- of the objective function value of the reoptimized solution $J_{\mathcal{I}^-}^{\text{reo}}$ compared to the objective function value of the initial solution $J_{\mathcal{I}}^*$ is non-negative, i. e.*

$$\Delta^- \geq 0. \tag{3.42}$$

Remark. *Since any task deletion reoptimization approach removes a task from the initial solution, Assumption 3.3 holds for the vast majority of reasonable task deletion reoptimization approaches.*

Furthermore, another assumption is relevant for giving an upper bound on the approximation ratio. Consider the modified problem instance $\mathcal{I}^-$ and a globally optimal solution $J_{\mathcal{I}^-}^*$ to be known. By adding task n_q to the problem instance $\mathcal{I}^-$, the initial problem instance $\mathcal{I}$ results. Solving this MRTA task insertion reoptimization problem (Problem 3.3) with $\mathcal{I}^-$ being the initial problem instance and $\mathcal{I}$ being the modified one, using any suitable reoptimization approach (e. g. the CMI defined in Algorithm 1), adds an increment Δ^+ in the objective function value to the initial value of the objective function, i. e.

$$J_{\mathcal{I}}^{\text{reo}} = J_{\mathcal{I}^-}^* + \Delta^+ \tag{3.43}$$

Building upon this relation, the following assumption is made:

Assumption 3.4. *Let the following hold for the initial problem instance $\mathcal{I}$, the modified task deletion problem instance $\mathcal{I}^-$ and the task deletion reoptimization approach: There exists a task insertion reoptimization approach such that the difference between the increment of the objective function Δ^+ of the task insertion reoptimization approach applied to $\mathcal{I}^-$ and the decrease in the objective function value Δ^- of the task deletion reoptimization approach applied to $\mathcal{I}$ does not exceed the optimal costs of the modified task deletion problem instance $\mathcal{I}^-$, i. e.*

$$\Delta^+ - \Delta^- \leq J_{\mathcal{I}^-}^*. \tag{3.44}$$

Numerous examples and problem instances investigated[27] have shown that Assumption 3.4 does not impose strong limitations and can be fulfilled for most problem

[27] These include the problem instances of the evaluation conducted in Chapter 5.

instances and reasonable task deletion approaches. For example, the CMI is a task insertion reoptimization approach that already fulfills $\Delta^+ \leq J^*_{\mathcal{I}^-}$ for most instances containing several tasks when adding task n_q to the modified task deletion problem instance $\mathcal{I}^-$. Together with Assumption 3.3, Assumption 3.4 is ensured for these cases.

Moreover, even if $\Delta^+ \geq J^*_{\mathcal{I}^-}$ holds and the worst case task insertion approximation ratio of $\alpha = 2$ is reached, it can be shown that if the task insertion reoptimization approach fulfills Assumption 3.2[28], as it is proven to be the case for the CMI by Lemma 3.7, Assumption 3.4 is fulfilled. In these cases it holds for the respective task insertion problem

$$\alpha = \frac{J^*_{\mathcal{I}^-} + \Delta^+}{J^*_{\mathcal{I}}} \overset{\text{Assumption 3.2}}{\leq} \frac{J^*_{\mathcal{I}^-} + \Delta^+}{\Delta^+} \overset{\Delta^+ \geq J^*_{\mathcal{I}^-}}{\leq} 2$$

Thus, $\Delta^+ = J^*_{\mathcal{I}^-} = J^*_{\mathcal{I}}$ must hold if an approximation ratio of $\alpha = 2$ shall result for the task insertion reoptimization problem for an instance in which $\Delta^+ \geq J^*_{\mathcal{I}^-}$ holds. Consequently, Assumption 3.4 is fulfilled in this case, too.

Under the previous assumptions, it is guaranteed by the following theorem that for any task deletion reoptimization approach that fulfills Assumptions 3.3 and 3.4 the resulting approximation ratio is bounded above.

Theorem 3.4

For every MRTA task deletion reoptimization approach solving an instance of Problem 3.4 such that Assumptions 3.3 and 3.4 hold, the resulting approximation ratio $\alpha = J^{\text{reo}}_{\mathcal{I}^-}/J^*_{\mathcal{I}^-}$ *is bounded above by* 2, *i. e.*

$$\alpha = \frac{J^{\text{reo}}_{\mathcal{I}^-}}{J^*_{\mathcal{I}^-}} \leq 2. \tag{3.45}$$

Proof:

The reoptimized solutions cannot have smaller objective function values than the globally optimal solutions to the respective problem instance. Together with (3.41) and (3.42),

$$J^*_{\mathcal{I}^-} \leq J^{\text{reo}}_{\mathcal{I}^-} \leq J^*_{\mathcal{I}} \leq J^{\text{reo}}_{\mathcal{I}} \tag{3.46}$$

follows. Using this relation together with (3.44) from Assumption 3.4 yields

$$\begin{aligned}
\alpha = \frac{J^{\text{reo}}_{\mathcal{I}^-}}{J^*_{\mathcal{I}^-}} &\overset{(3.41)}{=} \frac{J^*_{\mathcal{I}} - \Delta^-}{J^*_{\mathcal{I}^-}} \\
&\overset{(3.46)}{\leq} \frac{J^{\text{reo}}_{\mathcal{I}} - \Delta^-}{J^*_{\mathcal{I}^-}} \overset{(3.43)}{=} \frac{J^*_{\mathcal{I}^-} + \Delta^+ - \Delta^-}{J^*_{\mathcal{I}^-}} = 1 + \frac{\Delta^+ - \Delta^-}{J^*_{\mathcal{I}^-}} \\
&\overset{(3.44)}{\leq} 2.
\end{aligned}$$

[28] Please note that for the consideration of inserting task n_q to problem instance $\mathcal{I}^-$, instance $\mathcal{I}$ corresponds to $\mathcal{I}^+$ in the notation used for the task insertion reoptimization problem. Thus, using the notation of the task deletion reoptimization problem, Assumption 3.2 is given by $\Delta^+ \leq J^*_{\mathcal{I}}$.

□

A specific heuristic for solving the MRTA task deletion reoptimization problem, the *task deletion heuristic* (TDH) is introduced in the following.

3.3.3 Task Deletion Heuristic

In this section, a heuristic solution called the *task deletion heuristic* (TDH) to solve the MRTA task deletion reoptimization problem (Problem 3.4) is introduced. The heuristic is derived from the deletion procedure for the TSP of Archetti et al. [ABS03]. It erases the deleted task n_q from the route of the agent it is assigned to by the routing $X_{\mathcal{I}}^*$ of the initial solution $\vec{X}_{\mathcal{I}}^*$.

In the following, the agent, task n_q is assigned to by the initial solution, is denoted as m^*, vertex precedent to task n_q within the initial solution is denoted by i^* and the vertex subsequent to task n_q is denoted by j^*, i. e. within the routing $X_{\mathcal{I}}^*$ of the initial solution $\vec{X}_{\mathcal{I}}^*$ it holds that $x_{i^*,q}^{m^*} = x_{q,j^*}^{m^*} = 1$. Using this notation, the TDH applied to MRTA task deletion problem instances according to Problem 3.3 is defined as given by the following algorithm.

Algorithm 2 Task deletion heuristic (TDH)

1: Let the initial solution $\vec{X}_{\mathcal{I}}^*$ with the routing $X_{\mathcal{I}}^*$ be given. Within the route of agent m^*, i. e. the agent task n_q is assigned to by the routing $X_{\mathcal{I}}^*$, remove $x_{i^*,q}^{m^*}$ and $x_{q,j^*}^{m^*}$ from the solution, i. e. remove the edges (i^*, n_q) and (n_q, j^*) connecting task n_q to its precedent and subsequent vertices i^* and j^* within the route of agent m^*.

2: Insert edge (i^*, j^*) to the route of agent m^*, i. e. set $x_{i^*,j^*}^{m^*} = 1$.

3: Solve optimization problem (3.5) to determine the complete TDH solution $\vec{X}_{\mathcal{I}\text{-}}^{\text{TDH}}$ corresponding to the routing $X_{\mathcal{I}\text{-}}^{\text{TDH}}$ generated by the first two steps.

The decrease in the objective function value Δ^{TDH} caused by the TDH includes the reduction in transition times Δ_{c^t}, the reduction of task execution times Δ_{c^e} in the amount of execution time needed by agent m^* to perform task n_q, as well as a potential waiting time deviation Δ_{c^w}, i. e.

$$J_{\mathcal{I}\text{-}}^{\text{TDH}} = J_{\mathcal{I}}^* - \Delta^{\text{TDH}} \tag{3.47}$$

$$\text{with} \quad \Delta^{\text{TDH}} = \underbrace{d_{i^*,q}^{m^*} + d_{q,j^*}^{m^*} - d_{i^*,j^*}^{m^*}}_{\Delta_{c^t}} + \gamma \underbrace{d_q^{m^*}}_{\Delta_{c^e}} + \epsilon \Delta_{c^w}. \tag{3.48}$$

In the following section, the feasibility of solutions generated by the application of the TDH is analyzed.

3.3.4 Feasibility of the Solutions Generated by the Task Deletion Heuristic

After having introduced the TSH to solve the task deletion reoptimization problem according to Problem 3.4, its performance w. r. t. the feasibility of solutions generated is analyzed. According to the following Lemma 3.8, the feasibility of a solution of the TDH is guaranteed.

Lemma 3.8 (Solution feasibility for TDH)
Let an MRTA task deletion reoptimization problem according to Problem 3.4 be given such that the solution set of the modified problem instance $\mathcal{I}^-$ is non-empty. The TDH (see Algorithm 2) always finds a sulution $\vec{X}_{\mathcal{I}^-}^{\mathrm{TDH}}$ for the modified problem instance $\mathcal{I}^-$ that is feasible according to Lemma 3.1.

Proof:
To prove the feasibility of the routing $X_{\mathcal{I}^-}^{\mathrm{TDH}}$ generated by the TDH, Lemma 3.2 is applied.

By definition, the initial solution $\vec{X}_{\mathcal{I}}^*$ and consequently also its routing $X_{\mathcal{I}}^*$ is feasible according to Problem 3.4. Since all agent-task allocations remain unaltered for the TDH routing $X_{\mathcal{I}^-}^{\mathrm{TDH}}$ compared to the initial routing $X_{\mathcal{I}}^*$, Aspect L3.2.1 is guaranteed to be fulfilled by $X_{\mathcal{I}^-}^{\mathrm{TDH}}$ for all tasks within the modified instance $\mathcal{I}^-$.

Furthermore, the feasibility of the initial routing for the inital problem instance guarantees the corresponding CSG $\tilde{G}_{\mathcal{I},X_{\mathcal{I}}^*}$ to be acyclic according to Aspect L3.2.2 of Lemma 3.2. The CSG of the TDH routing $\tilde{G}_{\mathcal{I}^-,X_{\mathcal{I}^-}^{\mathrm{TDH}}}$ differs from $\tilde{G}_{\mathcal{I},X_{\mathcal{I}}^*}$ by three aspects:

1. In contrast to the CSP of the initial routing $\tilde{G}_{\mathcal{I},X_{\mathcal{I}}^*}$, the CSG of the TDH routing $\tilde{G}_{\mathcal{I}^-,X_{\mathcal{I}^-}^{\mathrm{TDH}}}$ does not contain the arcs corresponding to the precedence constraints related to task n_q that are included in the initial instance. Furthermore, it does not contain the arcs corresponding to the routing $x_{i^*,q}^{m^*}$ and $x_{q,j^*}^{m^*}$ which are both equal to one in the initial routing, but are set to zero in the TDH routing. However, deleting arcs from an acyclic graph cannot add any cycles to the graph.
2. The vertices of the CSG of the TDH routing $\tilde{G}_{\mathcal{I}^-,X_{\mathcal{I}^-}^{\mathrm{TDH}}}$ differ from the ones of the CSG of the initial routing $\tilde{G}_{\mathcal{I},X_{\mathcal{I}}^*}$ by no longer containing task n_q:
   - If synchronization constraints for task n_q are part of the initial instance $\mathcal{I}$, the synchronized vertex $\tilde{v}_q$ that contained task n_q in the initial CSG does no longer contain task n_q in the CSG of the TDH solution. This does not add a cycle to the CSG.

- If the task n_q is not constrained by any synchronization constraint in the initial instance $\mathcal{I}$, $\tilde{v}_q$ is deleted from the CSG of the initial instance. Since in this case, all arcs to $\tilde{v}_q$ are also deleted, the graph remains acyclic.

3. The arc $(\tilde{v}_{i^*}, \tilde{v}_{j^*})$ from the synchronized vertex $\tilde{v}_{i^*}$ to the synchronized vertex $\tilde{v}_{j^*}$ is added to the graph. No other arc is added. The CSG of the initial routing $\tilde{G}_{\mathcal{I},X^*_{\mathcal{I}}}$ is ensured to be acyclic and also contains a directed connection from $\tilde{v}_{i^*}$ to $\tilde{v}_{j^*}$ via the arcs $(\tilde{v}_{i^*}, \tilde{v}_q)$ and $(\tilde{v}_q, \tilde{v}_{j^*})$. Consequently, the insertion of the arc $(\tilde{v}_{i^*}, \tilde{v}_{j^*})$ cannot cause a cycle in the graph $\tilde{G}_{\mathcal{I}^-,X^{\mathrm{TDH}}_{\mathcal{I}^-}}$.

Moreover, the deletion of task n_q from the initial routing does not influence the start and end vertices of the agents' routes. Together with the feasibility of the initial routing, the fulfillment of Aspect L3.2.3 follows.

Consequently, all aspects of Lemma 3.2 are ensured to be fulfilled for the TDH routing $X^{\mathrm{TDH}}_{\mathcal{I}^-}$ and thus, $X^{\mathrm{TDH}}_{\mathcal{I}^-}$ is guaranteed to be feasible according to Lemma 3.1.

According to Corollary 3.1, completing the solution $\vec{X}^{\mathrm{TDH}}_{\mathcal{I}^-}$ by solving optimization problem (3.5) yields a feasible solution. □

An analysis of the approximation ratios resulting from the application of the TDS is presented in the following.

3.3.5 Upper Bounds on the Approximation Ratio for the Task Deletion Heuristic

As presented in the previous section, a solution generated by the TDH (see Algorithm 2) is ensured to be feasible (see Lemma 3.8). In this section, performance guarantees in the form of upper bounds on the approximation ratios are analyzed.

Analysis of the fulfillment of Assumptions 3.3 and 3.4

According to Theorem 3.4, the approximation ratio resulting from the application of the TDH is bounded above by $\alpha \leq 2$ for every task deletion reoptimization instance in which the TDH fulfills Assumptions 3.3 and 3.4. As analyzed previously, Assumption 3.4 is fulfilled for a vast majority of problem instances and can be guaranteed to be fulfilled even for problem instances in which a reoptimized solution to the respective task insertion problem would yield the worst possible approximation ratio of $\alpha = 2$.

The following lemma focuses on sufficient conditions for Assumption 3.3.

Lemma 3.9

If

1. *the weights γ, ϵ of the objective function* (3.1) *of the underlying optimization Problem 3.1 of the MRTA task deletion reoptimization Problem 3.3 fulfill* (3.7) *(i. e. $\epsilon \leq \min\{1, \gamma\}$) or*

2. *within the MRTA task deletion reoptimization problem instance (see Problem 3.4) neither precedence nor synchronization constraints must be considered (i. e. the instance is of configuration P_{het} or P_{hom}),*

Assumption 3.3 is fulfilled.

Proof:

The proof can be done analogously to the proof of Lemma 3.4 w. r. t. deleting a task. □

Remark. *The conditions given in Lemma 3.9 are sufficient. Hence, Assumption 3.3 still holds in many cases that do not fulfill either of the conditions. For example, positive waiting time deviations, i. e. the total amount of waiting time within the reoptimized solution $\vec{X}_{\mathcal{I}^-}^{TDH}$ being smaller than the sum of waiting times within the initial solution $\vec{X}_{\mathcal{I}}^*$, might occur. In these cases, the reduction of waiting time is caused by agents having to wait less time due to the task following the deleted task n_q in the initial solution (and potentially following tasks) being reached earlier in the modified solution.*

In summary, an approximation ratio of $\alpha = J_{\mathcal{I}^-}^{\text{TDH}} / J_{\mathcal{I}^-}^* \leq 2$ can be guaranteed for the vast majority of task deletion reoptimization instances since Assumptions 3.3 and 3.4 are most commonly fulfilled by the TDH. With the following proposition, the approximation ratio of $\alpha \leq 2$ is furthermore proven to be a tight bound for these instances.

Proposition 3.4

For the application of the TDH as defined in Algorithm 2 to instances of the MRTA task deletion reoptimization problem (Problem 3.4) for which Assumptions 3.3 and 3.4 are fulfilled, the bound on the approximation ratio of $\alpha \leq 2$ is tight, i. e. no lower upper bound on the approximation ratio exists.

Proof:

To prove that $\alpha \leq 2$ is the smallest upper bound possible on the approximation ratio resulting from the application of the TDH to MRTA task deletion reoptimization instances in which Assumptions 3.3 and 3.4 are fulfilled, an exemplay instance that fulfills both assumptions and for which the approximation ratio converges towards $\alpha \to 2$ for $\eta \to 0$, $\eta \geq 0$, for the application of the TDH is depicted in Figure 3.7. In the initial problem instance two agents with different depots have to execute two tasks.

The agents differ in their transition velocity, i. e. $v_1 = 1/2$ and $v_2 = 1$. Both agents are fully capable to execute both tasks, i. e. $a_1^1 = a_1^2 = 1$ and $a_2^1 = a_2^2 = 1$. For any $\eta > 0$, the globally optimal solution of the initial problem instance is unambigous and is given by assigning both tasks to agent k_2 which results in an objective function value of $J_{\mathcal{I}}^* = 4 - 2\eta$. In the modified problem instance, task n_2 is deleted. The resulting TDH solution has an objective function value of $J_{\mathcal{I}^-}^{\text{TDH}} = 4 - 2\eta$. The total deviation in objective function value thus equals $\Delta^{\text{TDH}} = J_{\mathcal{I}^-}^{\text{TDH}} - J_{\mathcal{I}}^* = 0$ and Assumption 3.3 is fulfilled. The globally optimal solution to the modified problem instance would be to assign task n_1 to agent k_1 which would yield an objective function value of $J_{\mathcal{I}^-}^* = 2$. Consequently, for $\eta \to 0$ the approximation ratio converges towards $\alpha \to 2$. Inserting the deleted task n_2 to the modified probem instance and using the CMI to insert n_2 to the globally optimal solution $J_{\mathcal{I}^-}^*$ would result in assigning task n_2 to agent k_2. Hence, it holds for the objective function differences of the task insertion and task deletion that $\Delta^+ - \Delta^- = \Delta^* - \Delta^{\text{TDH}} = 2 - 2\eta \leq 2 = J_{\mathcal{I}^-}^*$, i. e. also Assumption 3.4 is fulfilled. □

After analyzing the upper bound on the approximation ratio of the TDH for the general case with problem instances of configuration $P_{\mathcal{P},\mathcal{S},\text{het}}$, in the following a smaller upper bound is given for homogeneous problem instances of configuration P_{hom} that furthermore have one common depot.

Analysis of temporally unconstrained MRTA task deletion reoptimization problems with one common depot

For MRTA task deletion reoptimization problems without temporal constraints and with homogeneous teams of agents which moreover all start and end their route at the same position, i. e. problems of configuration P_{hom} for which additionally $o_1 = o_2 = \cdots = o_K := o_{\mathcal{K}}$ holds, an upper bound of $\alpha \leq 3/2$ can be guaranteed for the TDH.

Theorem 3.5

Solving an MRTA task deletion reoptimization problem (Problem 3.4) of configuration P_{hom} (see Table 3.2), for which additionally $o_1 = o_2 = \cdots = o_K := o_{\mathcal{K}}$ holds, by applying the TDH as defined in Algorithm 2, leads to an approximation ratio

$$P_{hom} \text{ with single depot:} \quad \alpha \leq \frac{3}{2}. \tag{3.49}$$

Proof:

This proof is inspired by the proof on the upper bound of the approximation ratio when deleting a vertex in the TSP by Archetti et al. [ABS03]. According to the problem

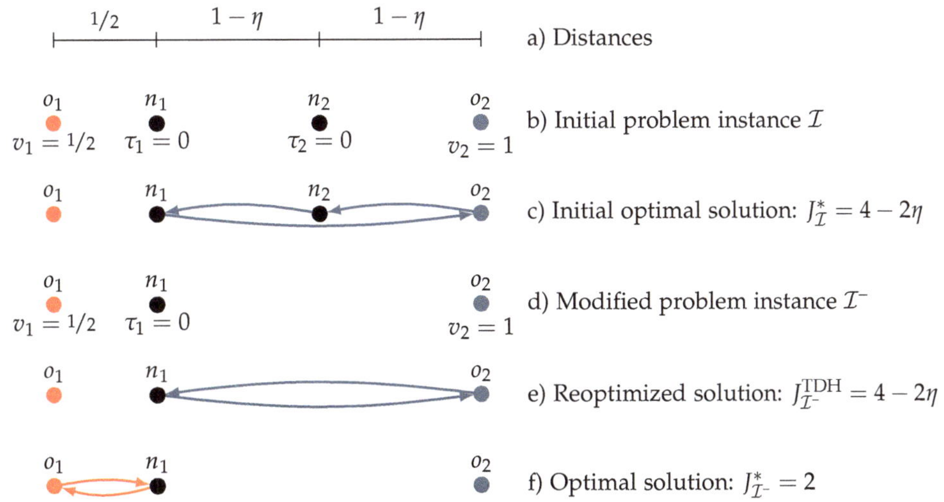

Figure 3.7: Example of an MRTA task deletion reoptimization problem instance with an approximation ratio of $\alpha \rightarrow 2$ for $\eta \rightarrow 0$ for the application of the TDH. The capabilities of both agents for both tasks are equal to one, i. e. $a_1^1 = a_1^2 = 1$ and $a_2^1 = a_2^2 = 1$. The figure is based on the author's publication [BKH+24].

configuration, all velocities as well as all capabilities are equal to one and no precedence and synchronization constraints are considered. The objective function value of the TDH solution is therefore given by

$$\begin{aligned} J_{\mathcal{I}^-}^{\mathrm{TDH}} &= J_{\mathcal{I}}^* - d_{i^*,q}^{m^*} - d_{q,j^*}^{m^*} + d_{i^*,j^*}^{m^*} - \gamma d_q^{m^*} \\ &\overset{v_{m^*}=a_q^{m^*}=1}{=} J_{\mathcal{I}}^* - \delta(i^*, n_q) - \delta(n_q, j^*) + \delta(i^*, j^*) - \gamma\tau_q. \end{aligned} \tag{3.50}$$

It furthermore holds that the value of the objective function of a globally optimal solution to the initial problem instance $\mathcal{I}$ is at least as high as the objective function value of the modified problem instance $\mathcal{I}^-$ augmented by inserting the task n_q to the route of any agent by tranistioning from any vertex to task n_q and back. Considering the velocities and capabilities all being equal to one, this relation is given by

$$J_{\mathcal{I}}^* \leq J_{\mathcal{I}^-}^* + \delta(i, n_q) + \delta(n_q, i) + \gamma\tau_q = J_{\mathcal{I}^-}^* + 2\delta(i, n_q) + \gamma\tau_q, \quad \forall i \in V. \tag{3.51}$$

Evaluating (3.51) for the precedent and subsequent vertex i^* and j^* of the deleted task n_q and adding it up yields

$$\begin{aligned} & 2J_{\mathcal{I}}^* \leq 2J_{\mathcal{I}^-}^* + 2\delta(i^*, n_q) + 2\delta(j^*, n_q) + 2\gamma\tau_q \\ \Leftrightarrow \quad & J_{\mathcal{I}}^* \leq J_{\mathcal{I}^-}^* + \delta(i^*, n_q) + \delta(j^*, n_q) + \gamma\tau_q \end{aligned} \tag{3.52}$$

By inserting (3.52) into (3.50)

$$J_{\mathcal{I}^-}^{\mathrm{TDH}} \leq J_{\mathcal{I}^-}^* + \delta(i^*, j^*) \tag{3.53}$$

follows. Since at least the vertices i^* and j^* are part of the modified problem instance, the agents share a common depot $o_{\mathcal{K}}$ and the triangle inequality holds for the distances, the following inequality holds for the optimal objective function value of the modified problem instance:

$$\begin{aligned} J^*_{\mathcal{I}^-} &\geq \delta(o_{\mathcal{K}}, i^*) + \delta(i^*, j^*) + \delta(o_{\mathcal{K}}, j^*) \\ &\geq 2\delta(i^*, j^*) \end{aligned} \tag{3.54}$$

Using (3.53) and (3.54), it follows for the approximation ratio

$$\begin{aligned} \alpha = \frac{J^{\mathrm{TDH}}_{\mathcal{I}^-}}{J^*_{\mathcal{I}^-}} &\overset{(3.53)}{\leq} \frac{J^*_{\mathcal{I}^-} + \delta(i^*, j^*)}{J^*_{\mathcal{I}^-}} = 1 + \frac{\delta(i^*, j^*)}{J^*_{\mathcal{I}^-}} \\ &\overset{(3.54)}{\leq} 1 + \frac{\delta(i^*, j^*)}{2\delta(i^*, j^*)} = \frac{3}{2}. \end{aligned} \tag{3.55}$$

□

Proposition 3.5
For the application of the TDH (Algorithm 2) to the task deletion reoptimization problem (Problem 3.4) of configuration P_{hom} (see Table 3.2), for which additionally $o_1 = o_2 = \cdots = o_K$ holds, the bound on the approximation ratio of $\alpha \leq \frac{3}{2}$ is tight, i. e. no lower upper bound on the approximation ratio exists.

Proof:
If only one agent is considered, the homogeneous MRTA task deletion reoptimization problem corresponds to a TSP. Furthermore, if only one agent is considered, the TDH corresponds to the vertex deletion heuristic applied by Archetti et al. [ABS03]. For the TSP, Archetti et al. [ABS03] prove the corresponding bound on the approximation ratio of $\alpha \leq 3/2$ to be tight [ABS03, Remark 2]. Since the TSP vertex deletion problem corresponds to a special case of the MRTA task deletion problem of configuration P_{hom} with one common depot, $\alpha \leq 3/2$ is a tight bound for the respective MRTA task deletion reoptimization problem. The problem instance proposed by Archetti et al. [ABS03] to prove the tight bound is depicted in Figure 3.8. □

In summary, the TDH (see Algorithm 2) is proven to always yield a feasible solution to the modified problem instance $\mathcal{I}^-$. Furthermore, it is shown to yield an approximation ratio bounded above by $\alpha \leq 2$ for the vast majority of problem instances. For problem instances of the configuration P_{hom} in which the agents share a common depot, a smaller and tight upper bound of $\alpha \leq 3/2$ is guaranteed.

In the following section, the reoptimization problem of varying the distances from one task to the other vertices within an MRTA problem instance is defined and analyzed.

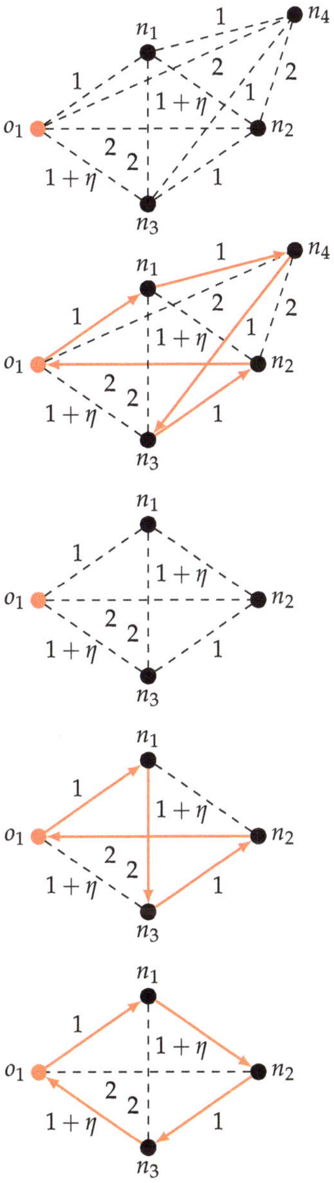

a) Initial problem instance $\mathcal{I}$ with distances and $v_1 = 1$, $\tau_i = 0, \forall i \in \mathcal{N}$

b) Initial optimal solution: $J^*_{\mathcal{I}} = 6$, $0 < \eta \leq 2$

c) Modified problem instance $\mathcal{I}^-$

d) Reoptimized solution: $J^{\text{TDH}}_{\mathcal{I}^-} = 6$

e) Optimal solution: $J^*_{\mathcal{I}^-} = 4 + 2\eta$, $0 < \eta < 1$

Figure 3.8: Example of an MRTA task deletion reoptimization problem instance of configuration P_{hom} with a common depot with an approximation ratio of $\alpha \to 3/2$ for $\eta \to 0$ for the application of the TDH (Example based on [ABS03, Fig. 3]). In this example one agent has to perform five tasks which all have a basic duration of $\tau_1 = \cdots = \tau_5 = 0$. Consequently, only transitioning times are represented in the objective function value.

3.4 Task Position Variation

This section investigates the reoptimization problem of varying a task position in an instance of a heterogeneous, time-extended MRTA problem with precedence and synchronization constraints. Two solution approaches are introduced and the feasibility of the solutions they generate is shown to be guaranteed.

3.4.1 Reoptimization Problem of Task Position Variation

The reoptimization problem of task position variation is, given a globally optimal solution $\vec{X}^*_{\mathcal{I}}$ to an initial problem instance $\mathcal{I}$, how to solve a related problem instance $\mathcal{I}^{\delta\updownarrow}$ in which the distances to and from one task n_q differ compared to the initial problem instance.

Problem 3.5 (MRTA reoptimization problem of task position variation)
The MRTA reoptimization problem of task position variation is an MRTA reoptimization problem according to Problem 3.2, where the modified problem instance $\mathcal{I}^{\delta\updownarrow}$ *is given by* $\mathcal{I}^{\delta\updownarrow} = \left\{\mathcal{N}, \mathcal{K}, \mathcal{O}, \mathcal{V}, \mathcal{A}, \mathcal{T}, \mathcal{D}^{\delta\updownarrow}, \mathcal{P}, \mathcal{S}\right\}$ *with the modified set*

- $\mathcal{D}^{\delta\updownarrow} = \{\delta(i,j), \delta(j,i) \in \mathcal{D} \,|\forall i,j \in \mathcal{N}, i,j \neq n_q\} \cup +$
$$\left\{\delta^{\updownarrow}(i,n_q), \delta^{\updownarrow}(n_q,i) \,|\forall i \in \mathcal{N}\right\}.$$

Thus, in the modified problem instance $\mathcal{I}^{\delta\updownarrow}$ *the distances to and from task* n_q *are different compared to the initial problem instance* $\mathcal{I}$.

Remark. *By definition of MRTA reoptimization problems (see Problem 3.2), the modified problem instance* $\mathcal{I}^{\delta\updownarrow}$ *is an MRTA problem instance according to Problem 3.1 which is why also the modified distances* $\mathcal{D}^{\delta\updownarrow}$ *are required to be metric.*

Remark. *If tasks and depots are locally distributed in a plane, the modification of varying the distances to one task corresponds to changing the position of the respective task and interpreting the distances between vertices* $\delta(i,j)$, $i,j \in V$, *as euclidean distances between their positions. The naming of this modification is based on this descriptive analogy.*

Two solution heuristics to the MRTA reoptimization problem of task position variation are given and analyzed w. r. t. their solutions' feasibility in the following sections. First, the simple yet efficient *initial solution approach* is introduced, followed by the more sophisticated *delete-insert heuristic*.

3.4.2 Initial Solution Approach

Given an instance of the MRTA reoptimization problem of task position variation (Problem 3.5), the *initial solution approach* (INI) is introduced in this section. Since INI can also be applied to different MRTA reoptimization problems resulting from other modifications (see Sections 3.5, 3.6, 3.7 and 3.9), it is introduced in a general manner without direct reference to the MRTA reoptimization problem of task position variation. The initial solution approach keeps the initial solution, as given in the following algorithm.

Algorithm 3 Initial solution approach (INI)

1: Given an globally optimal solution $X^*_{\mathcal{I}}$ and a modified problem instance $\mathcal{I}_{\text{mod}}$, set the routing $X^{\text{INI}}_{\mathcal{I}_{\text{mod}}}$ of the reoptimized solution $\vec{X}^{\text{INI}}_{\mathcal{I}_{\text{mod}}}$ to be equal to the routing $X^*_{\mathcal{I}}$ of the initial solution $\vec{X}^*_{\mathcal{I}}$, i. e.

$$X^{\text{INI}}_{\mathcal{I}_{\text{mod}}} = X^*_{\mathcal{I}}. \tag{3.56}$$

2: Solve optimization problem (3.5) corresponding to routing $X^{\text{INI}}_{\mathcal{I}_{\text{mod}}}$ to determine the complete the solution $\vec{X}^{\text{INI}}_{\mathcal{I}_{\text{mod}}}$.

To denote the specific modification INI is applied to, the subscript "$\mathcal{I}_{\text{mod}}$" is replaced by the respective reoptimization problem instance notation under consideration. Thus, $\vec{X}^{\text{INI}}_{\mathcal{I}^{\delta\updownarrow}}$ denotes the solution generated by INI to an MRTA reoptimization problem of task position variation.

For the reoptimization problem of task position variation, INI generates a feasible solution as shown in the following section.

3.4.3 Feasibility of the Solutions Generated by the Initial Solution Approach

The feasibility of the solutions generated by the application of INI to solve an MRTA reoptimization problem of task position variation according to Problem 3.5 is analyzed in this section. According to Lemma 3.10, the feasibility of the solution is guaranteed.

Lemma 3.10 (Solution feasibility of INI for the MRTA task position variation reoptimization probem)

Let an MRTA reoptimization problem of task position variation according to Problem 3.5 be given. INI (see Algorithm 3) always finds a solution $\vec{X}^{INI}_{\mathcal{I}^{\delta\updownarrow}}$ for the modified problem instance $\mathcal{I}^{\delta\updownarrow}$ that is feasible according to Lemma 3.1.

Proof:

To prove the feasibility of the solution $\vec{X}^{\mathrm{INI}}_{\mathcal{I}^{\delta\updownarrow}}$ generated by INI, Lemma 3.2 is applied. By definition, the initial solution $\vec{X}^*_{\mathcal{I}}$ is feasible for the initial problem instance $\mathcal{I}$ according to Problem 3.5. Since all agent-task allocations remain unaltered for the routing $X^{\mathrm{INI}}_{\mathcal{I}^{\delta\updownarrow}}$ of the INI solution $\vec{X}^{\mathrm{INI}}_{\mathcal{I}^{\delta\updownarrow}}$ compared to the routing $X^*_{\mathcal{I}}$ of the initial solution $\vec{X}^*_{\mathcal{I}}$, and also the agent capabilities $\mathcal{A}$ are equal in both problem instances $\mathcal{I}$ and $\mathcal{I}^{\delta\updownarrow}$, Aspect L3.2.1 is guaranteed to be fulfilled by the routing $X^{\mathrm{INI}}_{\mathcal{I}^{\delta\updownarrow}}$ for all tasks within the modified instance $\mathcal{I}^{\delta\updownarrow}$.

Furthermore, the feasibility of the initial solution for the inital problem instance guarantees the corresponding CSG $\tilde{G}_{\mathcal{I},X^*_{\mathcal{I}}}$ to be acyclic according to Aspect L3.2.2 of Lemma 3.2. Since neither does any agent-task allocation change, nor are any routing arcs, precedence arcs or synchronization constraints added, the CSG of the INI solution $\tilde{G}_{\mathcal{I}^{\delta\updownarrow},X^{\mathrm{INI}}_{\mathcal{I}^{\delta\updownarrow}}}$ does not differ from $\tilde{G}_{\mathcal{I},X^*_{\mathcal{I}}}$. Consequently, also Aspect L3.2.2 is ensured to remain fulfilled.

Moreover, since all agents' routes are unaltered compared to the initial routing, Aspect L3.2.3 must also remain fulfilled for the routing generated by INI. Thus, the INI routing $X^{\mathrm{INI}}_{\mathcal{I}^{\delta\updownarrow}}$ fulfills all aspects of Lemma 3.2 and is feasible.

According to Corollary 3.1, completing the solution $\vec{X}^{\mathrm{INI}}_{\mathcal{I}^{\delta\updownarrow}}$ by solving optimization Problem (3.5) yields a feasible solution. □

Remark. *By definition of Problem 3.5, the initial problem instance $\mathcal{I}$ is required to have a non-empty solution set as a globally optimal initial solution $\vec{X}_{\mathcal{I}}$ is given. Consequently, the modified problem instance $\mathcal{I}^{\delta\updownarrow}$ has a non-empty solution set, as well. This is because modifying the distances does not influence the feasibility of the routings corresponding to the solutions in the initial solution set, since, according to Lemma 3.2, routing feasibility does not depend on the distances. As given by Lemma 3.3, a feasible solution can always be found given a feasible routing. Thus, the solution set of the modified problem instance $\mathcal{I}^{\delta\updownarrow}$ is always non-empty, which is why INI is guaranteed to always find a feasible solution $\vec{X}^{INI}_{\mathcal{I}^{\delta\updownarrow}}$.*

3.4.4 Delete-Insert Heuristic

The second approach to solve an MRTA reoptimization problem of task position variation (Problem 3.5), called *delete-insert heuristic* (DIH), is introduced in this section. Similar to the INI heuristic introduced in Section 3.4.2, the DIH can also be applied to several MRTA reoptimization problems (see Section 3.5), which is why it is introduced in a general manner without direct reference to the MRTA reoptimization problem of task position variation.

The general idea of the DIH is to delete a task from an initial solution $\vec{X}_{\mathcal{I}}$ using the TDH and then to reinsert it using the CMI. However, if the task is associated with precedence or synchronization constraints, the CMI is not suited to insert a constrained task. Thus, an extension to the CMI, called *extended cheapest maximum insertion cost heuristic* (eCMI) is also introduced.

The eCMI can be applied to add a task n_{N+1} to a solution $\vec{X}_{\mathcal{I}}$ of an MRTA problem instance $\mathcal{I}$. When the task to be added n_{N+1} is constrained by precedence and/or synchronization constraints, it is not ensured that inserting the additional task on all edges of the routing $X_{\mathcal{I}}$ of solution $\vec{X}_{\mathcal{I}}$ yields a feasible routing. For example, the task n_{N+1} cannot be inserted on a precedent edge of a task, that is required to be a precedent task of n_{N+1}. Consequently, before evaluating a potential increase in objective function value, the feasibility of the routing resulting from the task insertion on the respective edge must be verified for each edge of the routing $X_{\mathcal{I}}$ by the application of Lemma 3.2. For this examination, the capability of the respective agent to perform task n_{N+1} is ensured and the CSG resulting from the task insertion on the respective edge is generated and checked for acyclicity. Since the vertices in which the agents start and end their routes are not altered by inserting task n_{N+1} on an edge of the initial solution, this aspect must not be verified (see routing feasibility criterion in Lemma 3.2).

For every edge for which the task insertion would yield a feasible routing, the maximum increase in objective function value resulting from this insertion, is evaluated. However, due to the additional precedence and/or synchronization constraints on the inserted task n_{N+1}, more waiting times might result than considered by the maximum insertion costs $\Delta^*_{\max}$ (see (3.13)). This happens for example, if the inserted task n_{N+1} is assigned such that it has to wait for precedent tasks to finish or for synchronized tasks to be ready to be executed. Also, other tasks related by a synchronization or precedence constraint to task n_{N+1} that have to spend extra waiting time for task n_{N+1} due to these constraints are not represented by the maximum insertion costs.

To overcome these limitations, the *extended maximum insertion costs* are introduced which also consider additional waiting times resulting from the precedence and synchronization constraints associated with the inserted task n_{N+1}. The extended maximum insertion costs also overapproximate the resulting waiting times by considering their maximum effect on all agents possibly affected by a temporal shift of tasks. For any feasible possible insertion edge, let $t_{N+1}(i, j, m)$ denote the earliest possible starting time of the inserted task n_{N+1} in the solution resulting from inserting it between vertices i and j in the route of agent m. Using this notation, the extended maximum insertion costs $\Delta^{i,j,m}_{\text{emax}}$ of inserting an additional task n_{N+1} with precedence and/or synchronization constraints between the vertices $i \in V$ and $j \in V$ on the route of agent $m \in \mathcal{K}$ in the solution $\vec{X}_{\mathcal{I}}$ are given by

$$\begin{aligned}\Delta^{i,j,m}_{\text{emax}} = \Delta^{i,j,m}_{\text{trans}} + \gamma d^{m}_{N+1} + \epsilon \Bigg\{ & \underbrace{\left(t_{N+1}(i,j,m) - t_i - d^m_i - d^m_{i,N+1}\right)}_{\text{waiting part A}} \\ & + \underbrace{\left(t_{N+1}(i,j,m) + d^m_{N+1} + d^m_{N+1,j} - t_i - d^m_i - d^m_{i,j}\right)\left(\beta_j - 1\right)}_{\text{waiting part B}} \\ & + \underbrace{\sum_{l \in \mathcal{N}} \max\left\{0, p_{N+1,l}\left(t_{N+1}(i,j,m) + d^m_{N+1} - t_l\right)\beta_l\right\}}_{\text{waiting part C}} \\ & + \underbrace{\sum_{l \in \mathcal{N}} \max\left\{0, s_{N+1,l}\left(t_{N+1}(i,j,m) - t_l\right)\beta_l\right\}}_{\text{waiting part D}} \Bigg\}. \end{aligned} \tag{3.57}$$

The extended maximum insertion costs comprise several parts. The first two summands describe the difference in transition and task execution times with $\Delta^{i,j,m}_{\text{trans}}$ being defined as given in (3.14). The other summands describe the maximum additional waiting times that might occur. Potential waiting times before the start of the execution of task n_{N+1} caused by precedence or synchronization constraints are considered by part A of the waiting times, which calculates the difference between the earliest possible starting time of task n_{N+1} without any waiting time and the earliest possible starting time $t_{N+1}(i,j,m)$ considering precedence and synchronization constraints. Waiting part B describes how much the insertion of task n_{N+1} between the vertices i and j in the route of agent m delays the start of vertex j compared to the initial solution $\vec{X}_{\mathcal{I}}$ without task n_{N+1}. As in the maximum insertion costs (see (3.13)), β_j describes the number of agents' routes that can be affected by a temporal shift of vertex j. Thus, the delay of vertex j can at most cause waiting times of the same amount as its own delay in $(\beta_j - 1)$ routes of other agents that need to wait for vertex j or other vertices following in the same route. The waiting part C describes the maximum waiting times caused by temporal shifts of tasks $l \in \mathcal{N}$ that by a precedence constraint are required to start their execution only after task n_{N+1} has been executed. In the same manner, waiting part D describes the maximum amount of waiting times caused by synchronization constraints of task n_{N+1} causing delays for other tasks $l \in \mathcal{N}$.

Remark. *The earliest possible starting time of a task n_{N+1} that is inserted between the vertices i and j in the route of agent m, depends on the time vertex i is finished and agent m is traversed to task n_{N+1}. It can possibly be delayed by one or more tasks $l \in \mathcal{N}$ having to be finished before the start of the execution of task n_{N+1} according to precedence constraints or by synchronized* *tasks $l \in \mathcal{N}$ that are ready for execution later that task n_{N+1}. Thus, the earliest possible starting time of task n_{N+1} is given by*

$$t_{N+1}(i,j,m) = \max \begin{cases} t_i + d_i^m + d_{i,N+1}^m \\ \max_{l \in \mathcal{N}} \left\{ p_{l,N+1} \sum_{m \in \mathcal{K}} \sum_{m \in \mathcal{N}} \left(t_l + d_l^m x_{l,m}^m \right) \right\}. \\ \max_{l \in \mathcal{N}} \left\{ s_{l,N+1} t_l \right\} \end{cases} \tag{3.58}$$

Using the extended maximum insertion costs, the eCMI applied to add a task n_{N+1} to a solution $\vec{X}_{\mathcal{I}}$ of an MRTA problem instance $\mathcal{I}$ is defined as given by the following algorithm.

Algorithm 4 Extended cheapest maximum insertion costs heuristic (eCMI)

1: **if** Task n_{N+1} is not related to any precedence or synchronization constraints **then**
2: Apply the CMI according to Algorithm 1.
3: **else**
4: For each edge (i,j) that is part of the route of any agent $m \in \mathcal{K}$ in the optimal solution $X_{\mathcal{I}}^*$ of the initial problem instanse, i. e. for all (i,j,m) : $\left(x_{i,j}^m \in X_{\mathcal{I}} \wedge x_{i,j}^m = 1 \right)$, verify if inserting task n_{N+1} on that edge yields a feasible routing according to Lemma 3.2, i. e.

   1. verity that $a_{N+1}^m > 0$ and
   2. ensure that the CSG resulting from the respective insertion is acyclic.

5: For every edge on which inserting task n_{N+1} yields a feasible routing according to step 4, calculate the extended maximum insertion costs $\Delta_{\text{emax}}^{i,j,m}$ according to (3.57).
6: To determine the reoptimized routing X^{eCMI}, insert task n_{N+1} in the route of agent $\hat{m}$ between the vertices $\hat{i}$ and $\hat{j}$ that correspond to the finite minimum of the above determined extended maximum insertion costs $\Delta_{\text{emax}}^{i,j,m}$. This gives the reoptimized routing X^{eCMI}. The resulting optimal extended maximum insertion cost Δ_{emax}^* is given by

$$\Delta_{\text{emax}}^* = \Delta_{\text{emax}}^{\hat{i},\hat{j},\hat{m}} = \min_{\left\{ (i,j,m) \middle| x_{i,j}^m \in X_{\mathcal{I}}^* \wedge x_{i,j}^m = 1 \right\}} \Delta_{\text{emax}}^{i,j,m}, \tag{3.59}$$

$$\Delta_{\text{emax}}^* \in \mathbb{R}_{\geq 0}. \tag{3.60}$$

7: Solve optimization problem (3.5) corresponding to routing X^{eCMI} to determine the complete solution $\vec{X}^{\text{eCMI}}$.
8: **end if**

Remark. *If the task to add n_{N+1} is associated with precedence and/or synchronization constraints, the eCMI verifies the feasibility of the generated solution. However, since only the solutions that can be generated by inserting task n_{N+1} to the initial solution $\vec{X}_{\mathcal{I}}$ are considered, it is not ensured that the eCMI always finds a feasible solution, even if feasible solutions to the modified problem instance exist.*

Remark. *Since the vertices in which the agents start and end their routes are not altered by inserting task n_{N+1} on an edge of the initial solution, this aspect of the routing feasibility (see Lemma 3.2) must not be verified as the initial solution is guaranteed to be feasible.*

Using the eCMI, the delete-insert heuristic applied to a task n_q within a solution $\vec{X}_{\mathcal{I}}$ of an MRTA problem instance $\mathcal{I}$ is defined as given by the following algorithm.

Algorithm 5 Delete-insert heuristic (DIH)

1: Apply the TDH as defined in Algorithm 2 to remove task n_q and all related precedence and synchronization constraints from solution $\vec{X}_{\mathcal{I}}$ to generate the intermediate solution $\vec{X}_{\mathcal{I}}^{-}$.
2: Insert task n_q considering all related precedence and synchronization constraints to the intermediate solution $\vec{X}_{\mathcal{I}}^{-}$ using the eCMI (see Algorithm 4).

For the MRTA reoptimization problem of task position variation, the DIH is ensured to generate a feasible solution as demonstrated in the following section.

3.4.5 Feasibility of the Solutions Generated by the Delete-Insert Heuristic

The feasibility of the solutions generated by the application of the DIH to solve an MRTA reoptimization problem of task position variation according to Problem 3.5 is analyzed in this section. According to Lemma 3.10, the DIH is guaranteed to find a feasible solution.

Lemma 3.11 (Solution feasibility of the DIH for the MRTA task position variation reoptimization probem)
Let an MRTA reoptimization problem of task position variation according to Problem 3.5 be given. The application of the DIH (see Algorithm 5) to task n_q always finds a solution $\vec{X}_{\mathcal{I}^{\delta\updownarrow}}^{DIH}$ for the modified problem instance $\mathcal{I}^{\delta\updownarrow}$ that is feasible according to Lemma 3.1.

Proof:
According to Lemma 3.8, the intermediate solution $\vec{X}_{\mathcal{I}}^{-}$ generated by the TDH in the first step of the DIH is guaranteed to be feasible. In the second step of the TDH, task n_q is reinserted using the CMI, if task n_q is not associated with any precedence or synchronization constraints. According to Lemma 3.5, the resulting solution is guaranteed to be feasible. If task n_q is constrained by precedence and/or synchronization constraints, steps 4 and 5 of the eCMI ensures that only feasible solutions can be generated. Since the initial solution $\vec{X}_{\mathcal{I}}$ always is a possible solution that can be generated by the TDH, which furthermore is ensured to be feasible for the modified problem instance $\mathcal{I}^{\delta\updownarrow}$ according to Lemma 3.10, the application of the DIH to task n_q is guaranteed to find a feasible solution to the modified problem instance $\mathcal{I}^{\delta\updownarrow}$. □

Remark. *As laid out in Section 3.4.3, the modified problem instance $\mathcal{I}^{\delta\updownarrow}$ (see Problem 3.5) always has a non-empty solution set such that INI always finds a feasible solution to the modified problem instance $\mathcal{I}^{\delta\updownarrow}$.*

Thus, both INI and DIH are suited to generate a feasible solution for the reoptimization problem of task position variation. INI has the potential to generate reoptimization solutions fast without considering alternative routing options. In contrast, DIH investigates several routing options with higher expected computation time. The modification of varying the duration of a task is investigated in the following section.

3.5 Task Duration Variation

The task duration variation reoptimization problem is introduced in this section and the feasibility of solutions generated by the previously introduced INI heuristic (see Algorithm 3) and by the DIH (see Algorithm 5) are analyzed.

3.5.1 Reoptimization Problem of Task Duration Variation

The definition of the MRTA reoptimization problem of task duration variation is given by Problem 3.6.

Problem 3.6 (MRTA reoptimization problem of task duration variation)
The MRTA reoptimization problem of task duration variation is an MRTA reoptimization problem according to Problem 3.2, where the modified problem instance $\mathcal{I}^{\tau^{\updownarrow}}$ *is given by* $\mathcal{I}^{\tau^{\updownarrow}} = \left\{\mathcal{N}, \mathcal{K}, \mathcal{O}, \mathcal{V}, \mathcal{A}, \mathcal{T}^{\tau^{\updownarrow}}, \mathcal{D}, \mathcal{P}, \mathcal{S}\right\}$ *with the modified set*

- $\mathcal{T}^{\tau^{\updownarrow}} = \{\tau_i \in \mathcal{T} \,|\, \forall i \in \mathcal{N}, i \neq n_q\} \cup \left\{\tau_q^{\updownarrow}\right\}$, *with* $\tau_q^{\updownarrow} \in \mathbb{R}_{\geq 0}$.

Thus, in the modified problem instance $\mathcal{I}^{\tau^{\updownarrow}}$*, the basic task duration of task* n_q *is different compared to the initial problem instance* $\mathcal{I}$.

To solve the MRTA reoptimization problem of task duration variation, INI and the application of DIH to task n_q with the modified duration are suitable. Both reoptimization approaches are analyzed w. r. t. the feasibility of the resulting solutions in the following.

3.5.2 Feasibility of the Solutions Generated by the Initial Solution Approach

The application of INI (see Algorithm 3) yields a feasible solution to the task duration variation reoptimization problem as given by Lemma 3.12.

Lemma 3.12 (Solution feasibility of INI for the MRTA task duration variation reoptimization probem)
Let an MRTA reoptimization problem of task duration variation according to Problem 3.6 be given. The application of INI (see Algorithm 3) always yields a solution $\vec{X}^{INI}_{\mathcal{I}^{\tau^{\updownarrow}}}$ *for the modified problem instance* $\mathcal{I}^{\tau^{\updownarrow}}$ *that is feasible according to Lemma 3.1.*

Proof:
The proof of the feasibility of the solution $\vec{X}^{\text{INI}}_{\mathcal{I}^{\tau^{\updownarrow}}}$ to the task duration variation problem instance $\mathcal{I}^{\tau^{\updownarrow}}$ is identical to the proof of Lemma 3.10. □

Remark. *The modified problem instance* $\mathcal{I}^{\tau^{\updownarrow}}$ *is ensured to always have a non-empty solution set, which is why INI is guaranteed to always find a feasible solution* $\vec{X}^{INI}_{\mathcal{I}^{\tau^{\updownarrow}}}$. *This is because modifying the task durations does not affect the feasibility of the corresponding routings since, according to Lemma 3.2, routing feasibility does not depend on task durations. By definition, the initial problem instance* $\mathcal{I}$ *has a non-empty solution set (see Problem 3.6). Together with*

Lemma 3.3 ensuring that feasible solutions can be determined for feasible routings, it follows that the solution-set of the modified problem instance $\mathcal{I}^{\tau^\updownarrow}$ in non-empty.

Also the DIH applied to task n_q is ensured to always yield a feasible solution as given by Lemma 3.13 as shown in the following section.

3.5.3 Feasibility of the Solutions Generated by the Delete-Insert Heuristic

Solving an MRTA reoptimization problem of task duration variation by the application of DIH (see Algorithm 5) always yields a feasible solution as given by Lemma 3.13.

Lemma 3.13 (Solution feasibility of the DIH for the MRTA task duration variation reoptimization probem)
Let an MRTA reoptimization problem of task duration variation according to Problem 3.6 be given. The application of the DIH (see Algorithm 5) to task n_q always finds a solution $\vec{X}^{DIH}_{\mathcal{I}^{\tau^\updownarrow}}$ for the modified problem instance $\mathcal{I}^{\tau^\updownarrow}$ that is feasible according to Lemma 3.1.

Proof:
Since INI is guaranteed to yield a feasible solution according to Lemma 3.12, the proof of Lemma 3.13 can be done analogously to the proof of Lemma 3.11. □

Remark. *As outlined in Section 3.5.2, the modified problem instance $\mathcal{I}^{\tau^\updownarrow}$ always has a non-empty solution set. Consequently, a feasible solution can always be found by the application of the DIH.*

As given by Lemmas 3.12 and 3.13, both INI and the DIH are suited to generate a feasible solution for the MRTA task duration variation problem.

In the following sections, the modifications of varying the capability or the velocity of an agent are investigated. For the corresponding reoptimization problems, INI is a suitable solution approach.

3.6 Agent Capability Variation

This section investigates variations in an agent's capabilities to perform tasks in the context of heterogeneous, time-extended MRTA problems with precedence and synchronization constraints. The corresponding reoptimization problem is defined and it is shown that the previously introduced INI heuristic (see Algorithm 3) always yields a feasible solution.

3.6.1 Reoptimization Problem of Agent Capability Variation

The definition of the MRTA reoptimization problem of agent capability variation is given by Problem 3.7.

Problem 3.7 (MRTA reoptimization problem of agent capability variation)
The MRTA reoptimization problem of agent capability variation is an MRTA reoptimization problem according to Problem 3.2, where the modified problem instance $\mathcal{I}^{a^{\updownarrow}}$ is given by $\mathcal{I}^{a^{\updownarrow}} = \left\{\mathcal{N}, \mathcal{K}, \mathcal{O}, \mathcal{V}, \mathcal{A}^{a^{\updownarrow}}, \mathcal{T}, \mathcal{D}, \mathcal{P}, \mathcal{S}\right\}$ with the modified set

- $\mathcal{A}^{a^{\updownarrow}} = \{a_i^m \in \mathcal{A} \,|\, \forall i \in \mathcal{N}, m \in \mathcal{K}, m \neq k_q\} \cup \left\{a_1^{q\updownarrow}, \ldots, a_N^{q\updownarrow}\right\}$
  $$\text{with } a_i^{q\updownarrow} \begin{cases} \in \mathbb{R}_{>0} & \text{if } a_i^q > 0 \text{ for } a_i^q \in \mathcal{A} \\ \in \mathbb{R}_{\geq 0} & \text{if } a_i^q = 0 \text{ for } a_i^q \in \mathcal{A} \end{cases}$$

Thus, in the modified problem instance $\mathcal{I}^{a^{\updownarrow}}$, the capabilities of agent k_q to perform at least one of the tasks $i \in \mathcal{N}$ are different compared to the initial problem instance $\mathcal{I}$.

To solve instances of the MRTA reoptimization problem of agent capability variation, INI can be applied. The feasibility of the solutions generated is analyzed in the following.

3.6.2 Feasibility of the Solutions Generated by the Initial Solution Approach

The application of the INI heuristic as defined by Algorithm 3 yields a feasible solution to the reoptimization problem of agent capability variation as given by Lemma 3.14.

Lemma 3.14 (Solution feasibility of INI for the MRTA reoptimization problem of agent capability variation probem)
Let an MRTA reoptimization problem of agent capability variation according to Problem 3.7 be given. The application of INI (see Algorithm 3) always yields a solution $\vec{X}^{\mathrm{INI}}_{\mathcal{I}^{a^{\updownarrow}}}$ for the modified problem instance $\mathcal{I}^{a^{\updownarrow}}$ that is feasible according to Lemma 3.1.

Proof:
The proof of the feasibility of the solution $\vec{X}^{\mathrm{INI}}_{\mathcal{I}^{a^{\updownarrow}}}$ to the agent capability variation problem instance $\mathcal{I}^{a^{\updownarrow}}$ is similar to the proof of Lemma 3.10. Since according to Problem 3.7 the capabilities of agent k_q are altered such that the agent is ensured to remain capable of executing all tasks it was capable of executing in the initial problem instance,

the task allocation defined by the initial routing remains feasible. No other aspect of the problem instance that could influence the feasibility of the initial routing for the modified problem instance is modified in problem instance $\mathcal{I}^{a\updownarrow}$ compared to the initial instance $\mathcal{I}$. Consequently, INI yields a routing $X^{\text{INI}}_{\mathcal{I}^{a\updownarrow}}$ that is feasible for the modified problem instance $\mathcal{I}^{a\updownarrow}$. According to Corollary 3.1, completing the solution $\vec{X}^{\text{INI}}_{\mathcal{I}^{a\updownarrow}}$ by solving optimization Problem (3.5) yields a feasible solution. □

Remark. *By definition of Problem 3.7, the initial problem instance $\mathcal{I}$ has a non-empty solution set. Since it is furthermore required that within the modified problem instance $\mathcal{I}^{a\updownarrow}$, agent k_q has a positive capability for all tasks for which its capabilities are positive in the initial problem instance $\mathcal{I}$, the solution set of the modified problem instance $\mathcal{I}^{a\updownarrow}$ is non-empty by definition. The requirement of positive capabilities remaining positive in the modified problem instance ensures that the routings of all solutions in the initial solution set remain feasible routings for the modified problem instance. This is due to the agent-task allocations remaining feasible and the other aspects of Lemma 3.2 being unaffected. According to Lemma 3.3, feasible solutions can be determined for feasible routings. Consequently, the solution set of $\mathcal{I}^{a\updownarrow}$ is non-empty by definition and INI finds a feasible solution to any MRTA reoptimization problem of agent capability variation.*

The reoptimization problem resulting from varying the transition velocity of an agent is discussed in the following section.

3.7 Agent Velocity Variation

The reoptimization problem resulting from varying the transition velocity of an agent is introduced. The INI heuristic (see Algorithm 3) is applied to generate a solution to the modified problem instance which is shown to yield a feasible solution.

3.7.1 Reoptimization Problem of Agent Velocity Variation

The MRTA reoptimization problem of agent velocity variation is defined as given in Problem 3.8.

Problem 3.8 (MRTA reoptimization problem of agent velocity variation)
The MRTA reoptimization problem of agent velocity variation is an MRTA reoptimization problem according to Problem 3.2, where the modified problem instance $\mathcal{I}^{v\updownarrow}$ is given by $\mathcal{I}^{v\updownarrow} = \left\{\mathcal{N}, \mathcal{K}, \mathcal{O}, \mathcal{V}^{v\updownarrow}, \mathcal{A}, \mathcal{T}, \mathcal{D}, \mathcal{P}, \mathcal{S}\right\}$ with the modified set

- $\mathcal{V}^{v\updownarrow} = \{v_m \in \mathcal{V} \,|\forall m \in \mathcal{K}, m \neq k_q\} \cup \left\{{v_q}^{\updownarrow}\right\}$ *with* ${v_q}^{\updownarrow} \in \mathbb{R}_{>0}$.

Thus, in the modified problem instance $\mathcal{I}^{v\updownarrow}$, the transition velocity of agent k_q is different compared to the initial problem instance $\mathcal{I}$.

INI can be applied to solve the MRTA reoptimization problem of agent velocity variation. An analysis w. r. t. the feasibility of the solutions generated is given in the following.

3.7.2 Feasibility of the Solutions Generated by the Initial Solution Approach

To solve the velocity variation reoptimization problem, INI (see Algorithm 3) is suitable. As given by Lemma 3.15, INI is guaranteed to always yield a feasible solution.

Lemma 3.15 (Solution feasibility of INI for the MRTA agent velocity variation reoptimization probem)
Let an MRTA reoptimization problem of agent velocity variation according to Problem 3.8 be given. The application of INI (see Algorithm 3) always yields a solution $\vec{X}^{INI}_{\mathcal{I}^{v\updownarrow}}$ for the modified problem instance $\mathcal{I}^{v\updownarrow}$ that is feasible according to Lemma 3.1.

Proof:
The proof of the feasibility of the solution $\vec{X}^{\mathrm{INI}}_{\mathcal{I}^{v\updownarrow}}$ to the agent velocity variation problem instance $\mathcal{I}^{v\updownarrow}$ can be done analogously to the proof of Lemma 3.10. □

Remark. *The modified velocity ${v_q}^{\updownarrow} > 0$ does not influence the feasibility of the routings of the solutions in the non-empty solution set of the initial problem instance, which remain feasible for the modified problem instance $\mathcal{I}^{v\updownarrow}$. Since according to Lemma 3.3 feasible solutions can be determined for feasible routings, the solution set of the modified problem instance $\mathcal{I}^{v\updownarrow}$ is guaranteed to be non-empty and INI finds a feasible solution to any MRTA reoptimization problem of agent velocity variation.*

Several modifications varying properties of tasks or agents have been introduced and analyzed so far. In the following sections, the insertion and deletion of temporal constraints are considered.

3.8 Insertion of a Precedence or Synchronization Constraint

In this section, the modifications of inserting an additional precedence or synchronization constraint to the problem instance are analyzed. To solve these two kinds of reoptimization problems, the *extended delete-insert heuristic* (eDIH) is introduced. The results generated by the eDIH are furthermore proven to be feasible.

3.8.1 Reoptimization Problems of Precedence or Synchronization Constraint Insertion

The corresponding MRTA reoptimization problems of precedence constraint insertion and of synchronization constraint insertion are given in Problems 3.9 and 3.10, respectively.

Problem 3.9 (MRTA reoptimization problem precedence constraint insertion)
The MRTA reoptimization problem of precedence constraint insertion is an MRTA reoptimization problem according to Problem 3.2, where the modified problem instance $\mathcal{I}^{p+}$ is given by $\mathcal{I}^{p+} = \{\mathcal{N}, \mathcal{K}, \mathcal{O}, \mathcal{V}, \mathcal{A}, \mathcal{T}, \mathcal{D}, \mathcal{P}^{p+}, \mathcal{S}\}$ with the modified set

- $\mathcal{P}^{p+} = \mathcal{P} \setminus \left\{p_{\bar{i},\bar{j}}\right\} \cup \left\{p^+_{\bar{i},\bar{j}}\right\}$
  with $p_{\bar{i},\bar{j}} = 0$, $p_{\bar{i},\bar{j}} \in \mathcal{P}$ and $p^+_{\bar{i},\bar{j}} = 1$, $p^+_{\bar{i},\bar{j}} \in \mathcal{P}^{p+}$.

Thus, in the modified problem instance $\mathcal{I}^{p+}$, one additional precedence constraint between task $\bar{i} \in \mathcal{N}$ and task $\bar{j} \in \mathcal{N}$ has to be considered compared to the initial problem instance $\mathcal{I}$.

Problem 3.10 (MRTA reoptimization problem of synchronization constraint insertion)
The MRTA reoptimization problem of synchronization constraint insertion is an MRTA reoptimization problem according to Problem 3.2, where the modified problem instance $\mathcal{I}^{s+}$ *is given by* $\mathcal{I}^{s+} = \{\mathcal{N}, \mathcal{K}, \mathcal{O}, \mathcal{V}, \mathcal{A}, \mathcal{T}, \mathcal{D}, \mathcal{P}, \mathcal{S}^{s+}\}$ *with the modified set*

- $\mathcal{S}^{s+} = \mathcal{S} \setminus \left\{s_{\tilde{i},\tilde{j}}\right\} \cup \left\{s^{+}_{\tilde{i},\tilde{j}}\right\}$
  with $s_{\tilde{i},\tilde{j}} = 0$, $s_{\tilde{i},\tilde{j}} \in \mathcal{S}$ *and* $s^{+}_{\tilde{i},\tilde{j}} = 1$, $s^{+}_{\tilde{i},\tilde{j}} \in \mathcal{S}^{s+}$.

Thus, in the modified problem instance $\mathcal{I}^{s+}$*, one additional synchronization constraint between task* $\tilde{i} \in \mathcal{N}$ *and task* $\tilde{j} \in \mathcal{N}$ *has to be considered compared to the initial problem instance* $\mathcal{I}$.

It is likely, that the additional precedence or synchronization constraint added to the problem instance is not respected by the initial solution, which is why INI is not suited, neither for the reoptimization problem of precedence constraint insertion nor for the reoptimization problem of synchronization constraint insertion. To generate a feasible solution, a reallocation of a task might be required in many cases, for which the DIH is suitable. However, any precedence or synchronization constraint added relates two tasks, and the DIH (see Algorithm 5) only reallocates one task. Since it is not predetermined, for which task the reallocation yields the better solution, an extension to the DIH called *extended delete-insert heuristic* is introduced.

3.8.2 Extended Delete-Insert Heuristic

The *extended delete-insert heuristic* (eDIH) builds upon the same idea as the DIH (see Algorithm 5), which compares the extended maximum insertion costs (see (3.57)) for all possible allocations of one task. Extending this, the eDIH takes into consideration the extended maximum insertion costs resulting from the individual reallocation of one of the two newly constrained tasks and chooses the allocation corresponding to the lowest overapproximated objective function value. The eDIH applied to two tasks n_q and n_l within a solution $\vec{X}_{\mathcal{I}}$ of an MRTA problem instance $\mathcal{I}$ is defined as given by the following algorithm.

The eDIH can be applied to the two tasks that are newly related by a precedence or synchronization constraint, to solve the reoptimization problem of precedence or synchronization constraint insertion. This approach is analyzed w. r. t. the feasibility of the solutions generated in the following section.

Algorithm 6 Extended delete-insert heuristic (eDIH)

```
1: for tasks n_q and n_l do
2:     Apply the TDH according to Algorithm 2 to remove the task and all related
       precedence and synchronization constraints from solution X⃗_I to generate the
       intermediate solution X⃗_I^-.
3:     for each edge of the routing X_I^- of the intermediate solution do
4:         Evaluate the feasibilty of inserting the task together with its related prece-
           dence and synchronization constraints on the respective edge as defined
           in step 4 of the eCMI.
5:     end for
6:     for each edge on which inserting the task yields a feasible routing do
7:         Calculate the extended maximum insertion costs Δ_emax^{i,j,m} according to (3.57).
8:     end for
9: end for
10: To determine the reoptimized routing X^eDIH, perform the task deletion and inser-
    tion of either tasks n_q or task n_l, such that it corresponds to the finite minimum of
    the determined extended maximum insertion costs Δ_emax^{i,j,m}.
11: Solve optimization problem (3.5) corresponding to routing X^eDIH to determine the
    complete solution X⃗^eDIH.
```

3.8.3 Feasibility of the Solutions Generated by the Extended Delete-Insert Heuristic

In this section, the feasibility of solutions generated by the eDIH for reoptimization problems of precedence or synchronization constraint insertion is analyzed. According to Lemma 3.16, an eDIH solution generated for a reoptimization problem of precedence constraint insertion is ensured to be feasible.

Lemma 3.16 (Solution feasibility of the eDIH for the MRTA reoptimization probem of precedence constraint insertion)
Let an MRTA reoptimization problem of precedence constraint insertion according to Problem 3.9 be given. A solution $X_{\mathcal{I}^{p+}}^{\text{eDIH}}$ *generated by the application of the eDIH (see Definition 6) applied to the two tasks* $\tilde{i}$ *and* $\tilde{j}$ *that are newly constrained by a precedence constraint in the modified problem instance* $\mathcal{I}^{p+}$*, is feasible according to Lemma 3.1 for the modified problem instance* $\mathcal{I}^{p+}$*.*

Proof:
By definition, the eDIH only evaluates routings in line 7 that were proven to be feasible for the modified problem instance $\mathcal{I}^{p+}$ in line 4. Since only evaluated routings are considered in the final choice of the reoptimized routing in step 10, the routing of the reoptimized solution $X_{\mathcal{I}^{p+}}^{\text{eDIH}}$ is guaranteed to be feasible for the modified problem instance $\mathcal{I}^{p+}$. From Corollary 3.1 the feasibility of the solution $\vec{X}_{\mathcal{I}^{p+}}^{\text{eDIH}}$ generated by

solving optimization Problem (3.5) for the feasible routing $X_{\mathcal{I}^{p+}}^{\mathrm{eDIH}}$ follows. □

Not only is the solution generated by the eDIH guaranteed to be feasible for reoptimization problem of precedence constraint insertions, but also for the reoptimization problem of synchronization constraint insertions as given by Lemma 3.17.

Lemma 3.17 (Solution feasibility of the eDIH for the MRTA synchronization constraint insertion reoptimization probem)
Let an MRTA reoptimization problem of synchronization constraint insertion according to Problem 3.10 be given. A a solution $X_{\mathcal{I}^{s+}}^{eDIH}$ to the modified problem instance $\mathcal{I}^{s+}$ generated by the application of the eDIH (see Definition 6) to the two tasks $\tilde{i}$ and $\tilde{j}$ that are newly constrained by a synchronization constraint in the modified problem instance $\mathcal{I}^{s+}$ is feasible according to Lemma 3.1.

Proof:
The proof of Lemma 3.17 follows analogously to the proof of Lemma 3.16. □

Remark. *Since only the solutions that can be generated by deleting and re-inserting one of the two the newly constrained tasks $\tilde{i}$ or task $\tilde{j}$ in the initial solution $\vec{X}_{\mathcal{I}}$ are considered, it is not ensured that the eDIH always finds a feasible solution, even if feasible solutions to the modified problem instance exist. However, if a solution is found by the eDIH, it is guaranteed to be feasible according to Lemmas 3.16 and 3.17.*

Besides inserting an additional precedence or synchronization constraint to an MRTA problem instance, also the deletion of a precedence or synchronization constraint is a possible modification. It is analyzed in the subsequent section.

3.9 Deletion of a Precedence or Synchronization Constraint

The last modifications investigated in this chapter are the modifications of deleting a precedence or synchronization constraint from an MRTA problem instance. The corresponding reoptimization problems are defined in the following and an analysis of the feasibility of the solutions generated by the INI heuristic is conducted.

3.9.1 Reoptimization Problems of a Precedence or Synchronization Constraint Deletion

The MRTA reoptimization problem of precedence constraint deletion is defined in Problem 3.11.

Problem 3.11 (MRTA reoptimization problem of precedence constraint deletion)

The MRTA reoptimization problem of precedence constraint deletion is an MRTA reoptimization problem according to Problem 3.2, where the modified problem instance $\mathcal{I}^{p^-}$ is given by $\mathcal{I}^{p^-} = \{\mathcal{N}, \mathcal{K}, \mathcal{O}, \mathcal{V}, \mathcal{A}, \mathcal{T}, \mathcal{D}, \mathcal{P}^{p^-}, \mathcal{S}\}$ with the modified set

- $\mathcal{P}^{p^-} = \mathcal{P} \setminus \left\{p_{\tilde{i},\tilde{j}}\right\} \cup \left\{p^-_{\tilde{i},\tilde{j}}\right\}$
with $p_{\tilde{i},\tilde{j}} = 1$, $p_{\tilde{i},\tilde{j}} \in \mathcal{P}$ *and* $p^-_{\tilde{i},\tilde{j}} = 0$, $p^-_{\tilde{i},\tilde{j}} \in \mathcal{P}^{p^-}$.

Thus, the precedence constraint between task $\tilde{i}$ and task $\tilde{j}$ that has to be considered in the initial problem instance $\mathcal{I}$, is ignored in the modified problem instance $\mathcal{I}^{p^-}$.

Similarly, the reoptimization problem of synchronization constraint deletion is given in Problem 3.12.

Problem 3.12 (MRTA reoptimization problem of synchronization constraint deletion)

The MRTA reoptimization problem of synchronization constraint deletion is an MRTA reoptimization problem according to Problem 3.2, where the modified problem instance $\mathcal{I}^{s^-}$ is given by $\mathcal{I}^{s^-} = \{\mathcal{N}, \mathcal{K}, \mathcal{O}, \mathcal{V}, \mathcal{A}, \mathcal{T}, \mathcal{D}, \mathcal{P}, \mathcal{S}^{s^-}\}$ with the modified set

- $\mathcal{S}^{s^-} = \mathcal{S} \setminus \left\{s_{\tilde{i},\tilde{j}}\right\} \cup \left\{s^-_{\tilde{i},\tilde{j}}\right\}$
with $s_{\tilde{i},\tilde{j}} = 1$, $s_{\tilde{i},\tilde{j}} \in \mathcal{S}$ *and* $s^-_{\tilde{i},\tilde{j}} = 0$, $s^-_{\tilde{i},\tilde{j}} \in \mathcal{S}^{s^-}$.

Thus, the synchronization constraint between task $\tilde{i}$ and task $\tilde{j}$, which equal to one in the initial problem instance $\mathcal{I}$, is deleted, i. e. set to zero, in the modified modified problem instance $\mathcal{I}^{s^-}$.

INI (see Algorithm 3) can be applied to solve the MRTA reoptimization problems of precedence or synchronization constraint deletion. An analysis w. r. t. the feasibility of the solutions generated is given in the following.

3.9.2 Feasibility of the Solutions Generated by the Initial Solution Approach

To solve the MRTA reoptimization problems of precedence or synchronization constraint deletion, INI (see Algorithm 3) is suited to generate feasible solutions. As given by the following Lemma 3.18, INI is guaranteed to always yield a feasible solution for any MRTA reoptimization problem of precedence constraint deletion.

> **Lemma 3.18 (Solution feasibility of INI for the MRTA reoptimization probem of precedence constraint deletion)**
> *Let an MRTA reoptimization problem of precedence constraint deletion according to Problem 3.11 be given. The INI heuristic (see Algorithm 3) always yields a solution $\vec{X}^{\text{INI}}_{\mathcal{I}^{\text{P-}}}$ to the modified problem instance $\mathcal{I}^{\text{P-}}$ that is feasible according to Lemma 3.1.*

Proof:
To prove the feasibility of the solution $\vec{X}^{\text{INI}}_{\mathcal{I}^{\text{P-}}}$ generated by INI, Lemma 3.2 is applied. By definition, the initial solution $\vec{X}^{*}_{\mathcal{I}}$ is feasible for the initial problem instance $\mathcal{I}$ (see Problem 3.11). Since all agent-task allocations remain unaltered for the routing $X^{\text{INI}}_{\mathcal{I}^{\text{P-}}}$ of the INI solution $\vec{X}^{\text{INI}}_{\mathcal{I}^{\text{P-}}}$ compared to the routing $X^{*}_{\mathcal{I}}$ of the initial solution $\vec{X}^{*}_{\mathcal{I}}$, and also the agent capabilities $\mathcal{A}$ are equal in both problem instances $\mathcal{I}$ and $\mathcal{I}^{\text{P-}}$, Aspect L3.2.1 is guaranteed to be fulfilled by the routing $X^{\text{INI}}_{\mathcal{I}^{\text{P-}}}$. Since the initial routing is not altered, Aspect L3.2.3, which requires the route of each agent to start and end at its individual depot, remains fulfilled as well. Furthermore, the feasibility of the initial solution for the inital problem instance guarantees the corresponding CSG $\tilde{G}_{\mathcal{I},X^{*}_{\mathcal{I}}}$ to be acyclic according to Aspect L3.2.2 of Lemma 3.2. The CSG of the INI solution only differs from the CSG of the initial solution by the removal of the arc corresponding to the deleted precedence constraint. Thus, also the CSG of the INI solution is ensured to be acyclic. Consequently, all aspects of Lemma 3.2 are fulfilled and the routing $X^{\text{INI}}_{\mathcal{I}^{\text{P-}}}$ is ensured to be feasible. According to Corollary 3.1, completing the solution $\vec{X}^{\text{INI}}_{\mathcal{I}^{\text{P-}}}$ by solving optimization Problem (3.5) yields a feasible solution. □

Remark. *As explained in the previous proof, removing a precedence constraint from a feasible routing does not influence the routing's feasibility. Consequently, the routings of the solutions contained in the solution set of the initial problem instance $\mathcal{I}$ remain feasible for the modified problem instance $\mathcal{I}^{\text{P-}}$. Since the solution set of the initial problem instance $\mathcal{I}$ is non-empty by definition, and furthermore it is ensured by Lemma 3.3 that feasible solutions can be determined for feasible routings, the solution set of the modified problem instance $\mathcal{I}^{\text{P-}}$ is guaranteed to be non-empty. Consequently, a feasible solution to an MRTA reoptimization problem of precedence constraint deletion can always be determined by the application of INI.*

Similarly, also for the reoptimization problem of synchronization constraint deletion, INI is proven to yield a feasible solution by Lemma 3.19.

Lemma 3.19 (Solution feasibility of INI for the MRTA reoptimization probem of synchronization constraint deletion)
Let an MRTA reoptimization problem of synchronization constraint deletion according to Problem 3.12 be given. The INI heuristic (see Algorithm 3) always yields a solution $\vec{X}^{\mathrm{INI}}_{\mathcal{I}^{\mathrm{s-}}}$ to the modified problem instance $\mathcal{I}^{\mathrm{s-}}$ that is feasible according to Definition 3.1.

Proof:
To prove the feasibility of the solution $\vec{X}^{\mathrm{INI}}_{\mathcal{I}^{\mathrm{p-}}}$ generated by INI, again Lemma 3.2 is applied. Analogously to the proof of Theorem 3.18, Aspects L3.2.1 and L3.2.3 are guaranteed to be fulfilled by the routing $X^{\mathrm{INI}}_{\mathcal{I}^{\mathrm{s-}}}$. Again, the CSG corresponding to the initial solution $\tilde{G}_{\mathcal{I},X^*_{\mathcal{I}}}$ is acyclic. The deletion of the synchronization constraint modifies the CSG of the modified solution $\tilde{G}_{\mathcal{I}^{\mathrm{s-}},X^{\mathrm{INI}}_{\mathcal{I}^{\mathrm{s-}}}}$ compared to the CSG of the initial solution $\tilde{G}_{\mathcal{I},X^*_{\mathcal{I}}}$ in such a way that the synchronized vertex containing task $\tilde{i}$ and task $\tilde{j}$ of the initial CSG $\tilde{G}_{\mathcal{I},X^*_{\mathcal{I}}}$ is split into two synchronized vertices in the reoptimized CSG $\tilde{G}_{\mathcal{I}^{\mathrm{s-}},X^{\mathrm{INI}}_{\mathcal{I}^{\mathrm{s-}}}}$. The initial solution being feasible implies that different agents are allocated to task $\tilde{i}$ and task $\tilde{j}$ in the initial solution and furthermore, that no precedence constraint exists between those tasks. Consequently, the splitting of the corresponding synchronized vertices results in the splitted vertices $\tilde{v}_{\tilde{i}}$ and $\tilde{v}_{\tilde{j}}$ not being connected by an arc in the CSG anymore. Since no additional arcs are added, the resulting CSG $\tilde{G}_{\mathcal{I}^{\mathrm{s-}},X^{\mathrm{INI}}_{\mathcal{I}^{\mathrm{s-}}}}$ is guaranteed to be acyclic and the corresponding routing $X^{\mathrm{INI}}_{\mathcal{I}^{\mathrm{p-}}}$ is ensured to be feasible. According to Corollary 3.1, completing the solution $\vec{X}^{\mathrm{INI}}_{\mathcal{I}^{\mathrm{p-}}}$ by solving optimization Problem (3.5) yields a feasible solution. □

Remark. *As explained in the previous proof, removing a synchronization constraint from a feasible routing does not influence the routing's feasibility. Following the same reasoning as in the previous remark, this ensures the solution set of the modified problem instance $\mathcal{I}^{\mathrm{s-}}$ to be non-empty. Therefore, a feasible solution to an MRTA reoptimization problem of synchronization constraint deletion can always be determined by the application of INI.*

3.10 Summarizing Remarks on the Introduced Reoptimization Heuristics

Reoptimization heuristics for MRTA reoptimization problems belonging to different modifications have been introduced in this chapter. They are the first reoptimization heuristics proposed for heterogeneous, time-extended MRTA problems with precedence and synchronization constraints. Additionally, they allow for solving MRTA reoptimization problems corresponding to previously unconsidered modifications such as task duration variation, agent velocity variation and the insertion or deletion of a precedence or synchronization constraint (see Table 2.3). For the first time, also guarantees on the solutions they generate are given. These guarantees are listed in Table 3.4. The CMI, TDH, INI and DIH heuristic are guaranteed to always find a feasible solution for the respective reoptimization problem. Also for the eDIH, the feasibility of the solution is ensured, if a solution to the corresponding precedence or synchronization constraint insertion reoptimization problem is found. Moreover, upper bounds on the approximation ratios have been given for the modifications and corresponding heuristics of adding or deleting a task from a heterogeneous, time-extended MRTA problem instance with precedence and synchronization constraints.

If more than one modification is applied at a time, these modifications can be split into the individual modifications and the corresponding reoptimization heuristics can be applied consecutively. For example, to handle the insertion of two tasks, the CMI can be applied consecutively for each of the two tasks. This approach does not influence the feasibility guarantees given for the individual reoptimization heuristics, i. e. the resulting solutions are guaranteed to be feasible. The guarantees given on the upper bounds of the approximation ratios for the modifications of task deletion and task insertion (see Sections 3.2 and 3.3) grow exponentially with repeated application. However, the order in which the individual modifications are adressed, may influence the objective function value of the resulting solution.[29] Thus, permuting the applied reoptimization heuristics and choosing the best result may be an approach to further optimize the reoptimization result.

With the goal of further improving the results of the reoptimization heuristics by searching a broader area of the solution space, the following chapter introduces a combination of the previously introduced reoptimization heuristics with metaheuristic solution approaches.

[29] This holds for any other reoptimization heuristic than INI, since INI does not alter the routing of the initial solution.

Table 3.4: Overview of analyzed modifications, applied heuristics and guaranteed results.

Modification	Heuristic	Existence of solution	Solution feasibility	Approximation ratio bounded
task insertion	CMI	✓	✓	✓
task deletion	TDH	✓	✓	✓
task position variation	INI	✓	✓	–
	DIH	✓	✓	–
task duration variation	INI	✓	✓	–
	DIH	✓	✓	–
agent capability variation	INI	✓	✓	–
agent velocity variation	INI	✓	✓	–
precedence constraint insertion	eDIH	–	✓	–
synchronization constr. insertion	eDIH	–	✓	–
precedence constraint deletion	INI	✓	✓	–
synchronization constr. deletion	INI	✓	✓	–

4 Metaheuristic Reoptimization of Time-Extended MRTA Problems

The second research question to be answered in this thesis is whether the solutions generated by the reoptimization heuristics introduced in Chapter 3 can possibly be improved by additionally applying metaheuristic approaches (see Section 2.6). Metaheuristics, which are usually defined independently of a specific optimization problem, aim at balancing intensification and diversification. Thus, they generate new solutions with at least some properties of already investigated (good) solutions and allow for a broader search of the solution space in order to potentially escape local optima (see Definition 2.9). The problem-specific reoptimization heuristics introduced in Chapter 3 are all guaranteed to yield feasible solutions. For the modifications of inserting and deleting a task to a time-extended MRTA problem instance, furthermore upper bounds on the resulting approximation ratio have been given. These performance guarantees can be given despite the low computational effort they can be expected to require. In order to combine the advantages of both approaches and to answer the second research question, a combination of the previously defined heuristics with metaheuristic solution approaches is introduced in this chapter. After introducing the general approach for combining the previously introduced reoptimization heuristics with metaheuristic solution methods in Section 4.1, a realization based on a genetic algorithm (GA) is given in Section 4.2. This is followed by the details on the implementation of the GA in Section 4.3.

4.1 General Reoptimization Framework

In this section, the general idea of combining metaheuristic solution approaches with reoptimization heuristics to generate a metaheuristic reoptimization framework is presented.

As introduced in Section 2.4, metaheuristics can be classified into trajectory-based methods, using a single solution that is modified throughout the optimization procedure, and population-based methods that require multiple candidate solutions at a time. An initialization step is required by all metaheuristic optimization approaches and it is usually conducted randomly [LLY20]. However, since all solutions generated during the execution of the metaheuristic depend to some extent on the initial solution or solutions, the initialization of metaheuristics is known to have a relevant influence on the algorithms' performance [LLY20, AEA$^+$23, SVJ23]. Bearing this in mind, the initialization of metaheuristics seems to be a good way to combine heuristic and metaheuristic approaches. The application of heuristics to generate initial solution(s) for metaheuristics has already been shown to lead to improved performance [GXCW20, MGDVF21, AAS22]. Hence, using the reoptimization heuristics introduced in Chapter 3 to initialize metaheuristic solution approaches for heterogeneous, time-extended MRTA problems with precedence and synchronization constraints yields a metaheuristic reoptimization framework for respective MRTA problems. An overview of the general procedure of a resulting metaheuristic reoptimization framework for heterogeneous, time-extended MRTA problems with precedence and synchronization constraints is depicted in Figure 4.1 The first step of the metaheuristic reoptimization is the application of a reoptimization heuristic that uses the information of the initial solution $\vec{X}^*_{\mathcal{I}}$ and the applied modification to generate a reoptimized solution $\vec{X}^{\text{reo,heur}}_{\mathcal{I}_{\text{mod}}}$ to the modified problem instance. Subsequently, a metaheuristic is applied to solve the modified problem instance $\mathcal{I}_{\text{mod}}$. The deployed metaheuristic is initialized using the heuristically generated reoptimized solution $\vec{X}^{\text{reo,heur}}_{\mathcal{I}_{\text{mod}}}$. In case of a trajectory-based metaheuristic, the heuristically reoptimized solution is utilized as initial solution for the metaheuristic solution method. In the case of a population-based metaheuristic, the heuristically reoptimized solution becomes part of the initial population. Subsequently, the metaheuristic optimization procedure is applied and the metaheuristic reoptimization solution $\vec{X}^{\text{reo,meta}}_{\mathcal{I}_{\text{mod}}}$ results. In comparison to solely applying the previously introduced reoptimization heuristics, this approach offers a broader search of the solution space which comes at the cost of additional calculation time required for the metaheuristic.

Any metaheuristic optimization approach suitable for heterogeneous, time-extended MRTA problems with precedence and synchronization constraints can be applied within this framework to generate a specific metaheuristic reoptimization method. If the applied metaheuristic guarantees to retain the best solution generated throughout the search process, the initialization using the heuristically reoptimized solution allows for an anytime interruption of the search process still yielding a feasible solution. This does not hold for a random initialization, since initially no feasible solution might be generated. Consequently, the heuristic initialization is beneficial w. r. t. calculation time. In contrast to randomly initialized metaheuristics, moreover also the performance guarantees on the respective reoptimization heuristics (see Chapter 3) remain valid.

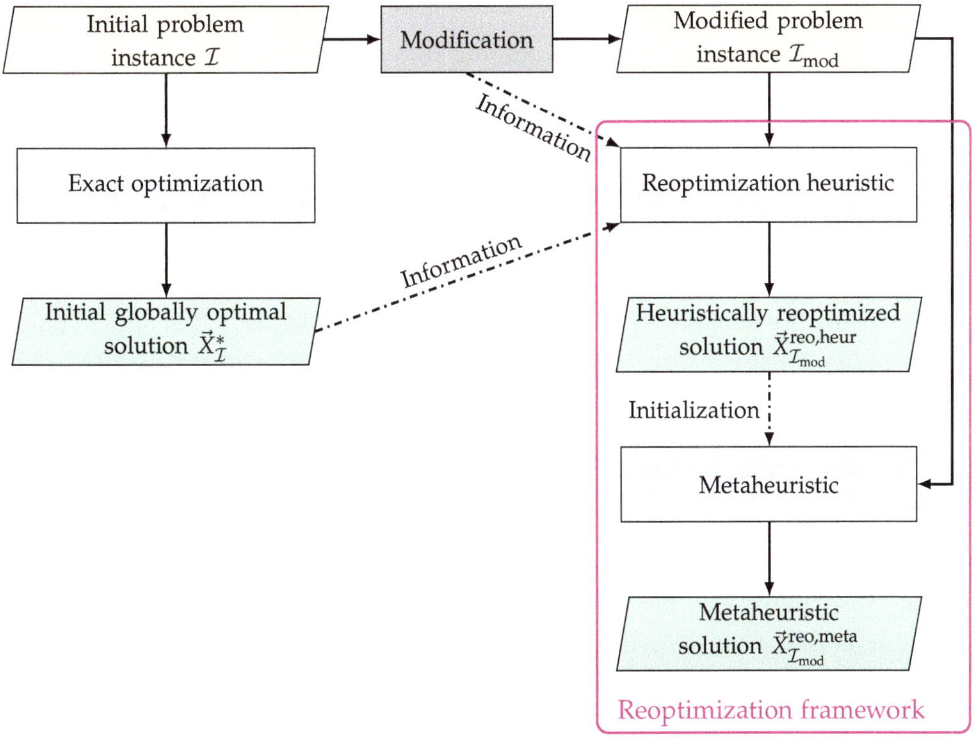

Figure 4.1: Overview of the metaheuristic reoptimization framework combining the reoptimization heuristics introduced in Chapter 3 with a metaheuristic solution approach.

Lemma 4.1 (Guarantees on the solutions generated by a metaheuristic MRTA reoptimization framework)
Let the specific metaheuristic applied in the metaheuristic reoptimization framework have the properties of

1. *yielding (as final solution) the best solution found throughout the metaheuristic search process and*
2. *infeasible solutions being evaluated worse than feasible solutions.*

Then, the application of this framework to any of the MRTA reoptimization problem defined in Chapter 3, i.e. Problems 3.3 to 3.12, ensures the guarantees on the feasibility of the applied heuristic solutions, i.e. Lemmas 3.5, 3.8 and 3.10 to 3.19, to remain valid for the solutions generated by the MRTA reoptimization framework. Furthermore, the upper bounds of the approximation rations for the CMI and TDH, i.e. Theorems 3.2 to 3.5, are preserved.

Proof:
The solutions generated by the heuristics introduced in Chapter 3 are utilized to initialize the metaheuristic search. These solutions are guaranteed to be feasible by Lemmas 3.5, 3.8 and 3.10 to 3.19. Consequently, for trajectory-based metaheuristics the initial solution is ensured to be feasible and for population-based metaheuristics at least one feasible solution is ensured to be contained in the initial population. Thus, if infeasible solutions are ensured to be evaluated worse than feasible solutions and the metaheuristic is defined such that it yields the best solution found throughout the search process, the guarantees given on the feasibility of the solutions of the reoptimization heuristics remain valid. The same argumentation holds for the upper bounds on the approximation ratio. □

The presented reoptimization framework allows for the combination of heuristic and metaheuristic solution approaches for heterogeneous, time extended MRTA reoptimization problems with precedence and synchronization constraints. If the respective reoptimization framework fulfills Lemma 4.1, the potential of making use of the advantages of both heuristic and metaheuristic approaches is given.

A realization of the introduced metaheuristic reoptimization framework using a genetic algorithm is presented in the following section.

4.2 MRTA Reoptimization Framework with Genetic Algorithm

In this section, a specific realization of the metaheuristic reoptimization framework is introduced.

The literature overview of approximative optimization methods for time-extended MRTA problems presented in Section 2.4 illustrates the great variety of metaheuristics applied to time-extended MRTA problems. However, the metaheuristic approach chosen for application in the reoptimization framework must be suitable to generate good results and to consider heterogeneous teams of robots, precedence as well as synchronization constraints. To find a metaheuristic approach that fulfills these requirements, an extensive investigation of solution approaches for time-extended MRTA problems, VRPs and related problems has been conducted in a bachelor thesis by Teufel [Teu20] that has been supervised by the author of this thesis. A GA-based approach for the HCCSP by Entezari and Mahootchi [EM20] was considered to best fulfill these requirements while at the same time allowing for application-specific modifications. The GA was implemented and adapted to the MRTA application. For the results presented in the bachelor thesis, an optimization considering different parameters, e. g. different population sizes, mutation and crossover probabilities, and different parent selection operators, was conducted to improve the performance of the respective GA on the heterogeneous, time-extended MRTA problems with precedence and synchronization constraints under consideration. This customized GA approach is combined with the reoptimization heuristics introduced in the previous Chapter 3 to generate a specific GA-based metaheuristic reoptimization framework. To evaluate the idea of combining heuristics with a GA for reoptimization, a preliminary examination with different heuristics without performance guarantees has been conducted. The promising results as well as the details of the implementation of the GA have been published by the author at an IEEE conference [BTIH21].

The general idea of GAs is to employ ideas originating from the theory of evolution, which leads to a special wording in this context [ZBB10]. A *population* describes a set of solutions. Within a GA, two solution representations, called *phenotype* and *genotype* are used. The phenotype describes a certain solution with all its features, i. e. a solution $\vec{X}_{\mathcal{I}}$ for an MRTA problem instance $\mathcal{I}$. The solution representation used within the evolutionary process is called genotype. Thus, in order to insert the heuristically generated reoptimization solution into the initial population of the GA, it is first translated into its genotype representation.[30] The *fitness* of an individual solution, which is usually determined using the phenotype representation, describes its quality [Kra17, Chapter 2] and is thus closely related to the objective function value given by (3.1).

In every iteration of the evolutionary cycle, solutions out of the current population are selected to become *parent solutions* which generate *offspring solutions*. A fitness-based *selection procedure* determines how the parent solutions are chosen. To converge towards optimal solutions, solutions with good fitness values should be preferred. However, it is reasonable for each solution to have a non-zero probability of being selected, since choosing suboptimal solutions out of the current population as parent solutions allows for overcoming local optima. To determine the offspring population, *crossover* and *mutation* are applied. The crossover operator implements a mechanism that mixes the features of the genotypes of (generally two) parents to generate new offspring solutions. With a certain probability, a mutation operator is applied to individual solutions, which changes a single genotype based on random alterations. A *replacement operator* determines which solutions out of previous population together with the offspring solutions generated in the current evolutionary cycle are chosen to form the new population for the next evolutionary cycle. The evolutionary loop is repeated until a termination condition, often given by a predefined number of iterations, is met. [Kra17, Chapter 2]

The corresponding procedure of the GA, including the initialization with the heuristically reoptimized solution, is depicted in Figure 4.2. The solution resulting from the GA-based reoptimization framework is denoted as $\vec{X}^{\text{reo,GA}}_{\mathcal{I}_{\text{mod}}}$. Details on the realization of the GA are given in the following Section.

30 Details on the genotype representation applied in the GA-based reoptimization framework are given in the following Section 4.3.

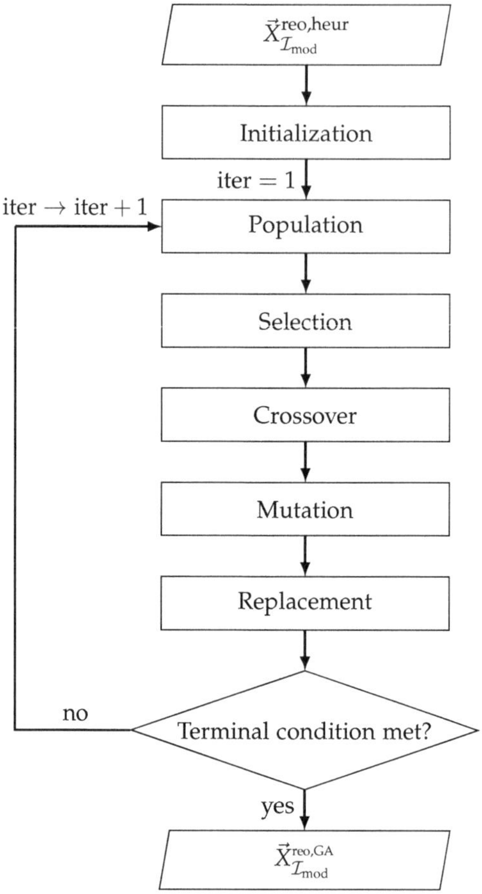

Figure 4.2: Overview of the evolutionary loop within the GA initialized with the solution of the reoptimization heuristic.

4.3 Genetic Algorithm Realization

After introducing the general idea of combining the reoptimization heuristics with a GA optimization and the GA evolutionary loop, details on the GA applied within this thesis for heterogeneous, time-extended MRTA problems with precedence and synchronization constraints are presented in this section. Some aspects are taken from the GA implementation for the HCCSP by Entezari and Mahootchi [EM20], while some aspects have been adapted to best fit the time-extended MRTA application under consideration in this thesis. This adaption, conducted for the bachelor thesis of Teufel [Teu20] which was supervised by the author of this thesis, is presented in the following.

Genotype representation

The genotype, i. e. the solution representation used within the GA, is defined equivalently as proposed by Entezari and Mahootchi [EM20]. The genotype consists of two matrices, an assignment matrix **A** and a scheduling matrix **S**. Both matrices are of dimension $1 \times N$. The assignment matrix **A** determines which task is executed by which agent. It contains the agent indices m with $\mathbf{A}(1,i) = m$ indicating that agent $k_m \in \mathcal{K}$ is assigned to task $n_i \in \mathcal{N}$. The order of task execution is contained within the scheduling matrix **S**. It contains a permutation of the task indices $i \in \{1, \ldots, N\}$ which defines the sequence in which the tasks are performed.

For example, let a problem instance $\mathcal{I}$ with three agents $\mathcal{K} = \{k_1, k_2, k_3\}$ and eight tasks $\mathcal{N} = \{n_1, \ldots, n_8\}$ be given. The genotype consisting of the two matrices

$$\mathbf{A} = \begin{bmatrix} 2 & 2 & 3 & 1 & 1 & 2 & 2 & 3 \end{bmatrix},$$
$$\mathbf{S} = \begin{bmatrix} 4 & 5 & 7 & 2 & 8 & 6 & 3 & 1 \end{bmatrix}$$

determines the routing $X_{\mathcal{I}}$ of the corresponding solution. It is set to agent k_1 conducting the route $\{o_1, n_4, n_5, o_1\}$, agent k_2 having the route $\{o_2, n_7, n_2, n_6, n_1, o_2\}$ and route $\{o_3, n_8, n_3, o_3\}$ being performed by agent k_3. Assuming the feasibility of the routing corresponding to a genotype, the timing information required for the complete solution representation is determined using (3.5).

Crossover and mutation operators

Apposite to the genotype representation, also the crossover and mutation operators are the same as proposed by Entezari and Mahootchi [EM20]. They are described in the following.

Both crossover operators for assignment and scheduling matrices generate two offspring matrices each. To crossover two assignment matrices, a randomly generated binary mask of the size $1 \times N$ determines which matrix entries are inherited from which parent to the offsprings. For one offspring, the entries of the first parent are taken for those entries in which the mask contains a 0 and the entries of the other parent are taken for entries in which the mask is equal to 1. This logic is vice versa for the other offspring. For the scheduling matrices, a single point crossover operator is chosen. It randomly chooses a point at which the parent chromosomes are split. For each offspring, the scheduling matrix before that point is equal to the parent. The remaining elements are filled up in the order they appear in the other parent. To illustrate these crossover operators, an example of the crossover operators applied to two parent genotypes $P1$ and $P2$ generating two offsprings $O1$ and $O2$ is depicted in Figure 4.3.

For the mutation of an assignment matrix **A**, a $(1 \times N)$-matrix with random entries uniformly distributed between 0 and 1 is generated. Given a predefined value $\kappa \in [0, 1]$,

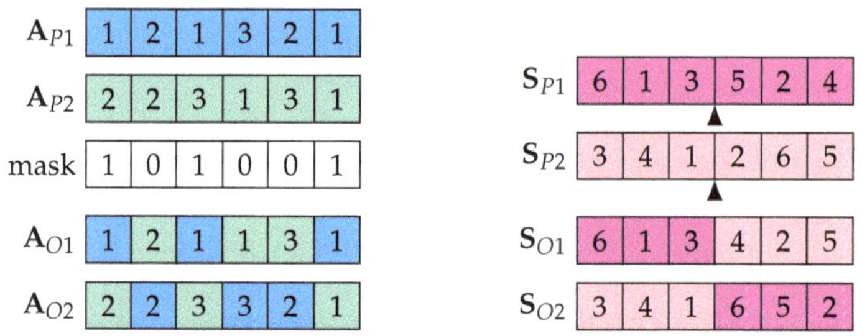

(a) Crossover of assignment matrices **A**.

(b) Crossover of scheduling matrices **S**.

Figure 4.3: Example of crossover operators on assignment and scheduling matrices. In this example, three agents and six tasks are considered. For the assignment matrix, a randomly generated binary mask determines for each offspring, which entry is inherited from which parent. For the crossover of the scheduling matrix, the randomly determined crossover point is marked by a black triangle. It determines until which point the order of the entries is inherited from which parent. The Figure is based on the author's publication [BTIH21] (©2021 IEEE).

all entries of the assignment matrix for which the value of the ramdomized matrix is greater or equal to κ remain unaltered while the others are modified. The modification is realized by randomly choosing an agent that is capable of performing the corresponding task. To mutate a scheduling matrix $\mathbf{S}$, one out of three two-point operators, i. e. swap, reversion and insertion, is chosen randomly. The swap operator exchanges the values of the two randomly chosen positions. The order of all entries between the two chosen positions is reversed, if the reversion operator is applied. The insertion operator deletes the entry at the second position and inserts it after the first position, while shifting all other entries in between one position to the right. An example of the mutation operators both for an assignment and a scheduling matrix is depicted in Figure 4.4.

Fitness evaluation

As in the publication of Entezari and Mahootchi [EM20], the goal of the GA optimization is to minimize fitness values, i. e. solutions with lower fitness values are preferred. Since Entezari and Mahootchi [EM20] consider a HCCSP, the fitness function has been adapted to suit the MRTA application.[31]

Since feasible solutions should be preferred over infeasible ones and furthermore no timing information can be determined for infeasible routings, the fitness determiniation differentiates between feasible and feasible solutions. As explained earlier, a genotype determines the routing $X_{\mathcal{I}}$ of a solution. For any genotype for which the corresponding routing is feasible according to Lemma 3.2, the missing timing information

[31] In the HCCSP considered by Entezari and Mahootchi [EM20], the objective function comprises aspects such as minimization of total overtime of staff members and continuity of care, which are not considered in the MRTA application. Furthermore, neither do Entezari and Mahootchi consider multiple depots nor agent-dependent task execution times.

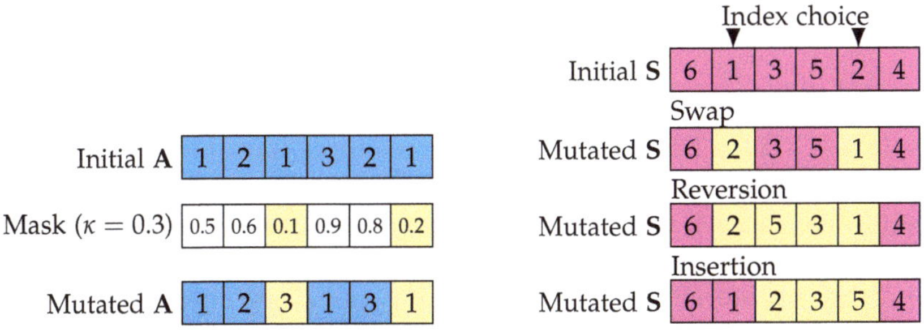

(a) Mutation of an assignment matrix **A**.

(b) Mutation of a scheduling matrix **S**.

Figure 4.4: Example of mutation operators on assignment and scheduling matrices. Entries highlighted in yellow indicate differences compared to the initial matrix. For the mutation of an assignment matrix, a randomly generated mask and value for κ determine that the assignments for all entries, having a smaller value than κ in the mask, are randomly changed. For the mutation of the scheduling matrix the black triangles symbolize the random choice of two indices. Depending on whether the swap, reversion of insertion operator are chosen, a different mutated scheduling matrix results. The Figure is based on the author's publication [BTIH21] (©2021 IEEE).

is determined using (3.5) to complete the solution, which, according to Corollary 3.1, yields a feasible solution. For any feasible solution, the fitness value is equivalent to the objective function value $J_{\mathcal{I}}$ (see (3.1)) of the solution.

If the genotypes corresponds to an infeasible routing however, no complete timing information can be determined. In these cases, a penalty function

$$f\left(X_{\mathcal{I}}\right) = c_{\text{stat}} + f_{L3.2.1}\left(X_{\mathcal{I}}\right) + f_{L3.2.2}\left(X_{\mathcal{I}}\right) \tag{4.1}$$

is applied to determine the fitness of the solution. A static penalty component is given by $c_{\text{stat}} \in \mathbb{R}_{>0}$, which is chosen such that it ensures the fitness values of infeasible solutions to be worse than the one of feasible solutions. In order to additionally include a measurement of the degree of infeasibility of a solution, the components $f_{L3.2.1}$ and $f_{L3.2.2}$ are introduced. They depend on the cause of the routing's infeasibility. If the assignments of tasks to robots are feasible, i. e. L3.2.1 of Lemma 3.2 is fulfilled, $f_{L3.2.1}$ equals zero. Otherwise, it is given by

$$f_{L3.2.1}(X_{\mathcal{I}}) = c_{\text{dyn}} n_{L3.2.1}(X_{\mathcal{I}}) \tag{4.2}$$

with $n_{L3.2.1}$ being the number of tasks that are assigned to an incapable robot and $c_{\text{dyn}} \in \mathbb{R}_{>0}$ being a constant greater zero. Similarly, $f_{L3.2.2}$ is equal to zero if the CSG corresponding to the routing $X_{\mathcal{I}}$ is acyclic, i. e. L3.2.2 of Lemma 3.2 is fulfilled. If this is not the case,

$$f_{L3.2.2}(X_{\mathcal{I}}) = c_{\text{dyn}} n_{L3.2.2}(X_{\mathcal{I}}) \tag{4.3}$$

determines the additional penalty factor with $n_{L3.2.2}$ being the number of cycles within the corresponding CSG. This fitness definition allows both feasible and infeasible solutions to be meaningfully considered within the evolutionary loop of the GA.

Selection procedure

The selection procedure used within this thesis differs from the one used by Entezari and Mahootchi [EM20]. Different selection procedures have been investigated and compared within the bachelor thesis of Teufel [Teu20]: *Tournament selection* showed the best performance w. r. t. solution fitness as well as solution feasibility. Tournament selection is a fitness-based selection operator. For the selection of each parent solution, a defined number of solutions is selected randomly out of the current population. The number of solutions selected is called tournament size. Out of the solution subset, the solution with the best fitness value is chosen to become a parent solution. [Kra17, Chapter 2.7]

Furthermore, the tournament size of solutions has been investigated and a size of three was found to yield the best results. An example of a tournament selection with a tournament size of three is depicted in Figure 4.5. Out of the initial population three solutions are randomly chosen and the best individual out of these three becomes the selected one.

Elitism

The GA proposed by Entezari and Mahootchi [EM20] applies a replacement strategy in which a population is fully replaced by the newly generated offspring population in each iteration. In the GA applied in this thesis, this strategy is extended by the principle of elitism [ES15, Chapter 5.3.2]: In every evolutionary loop, the amount of offspring solutions generated equals the population size. In order to prevent the loss of the best solution out of the previous population, the best solution of the previous population is migrated into the new population and replaces the worst offspring generated in the current evolutionary loop. An example of the principle of elitism is depicted in Figure 4.6.

Due to the application of the principle of elitism together with the solutions' fitness evaluation, Lemma 4.1 applies to this specification of the GA-based MRTA reoptimization framework.

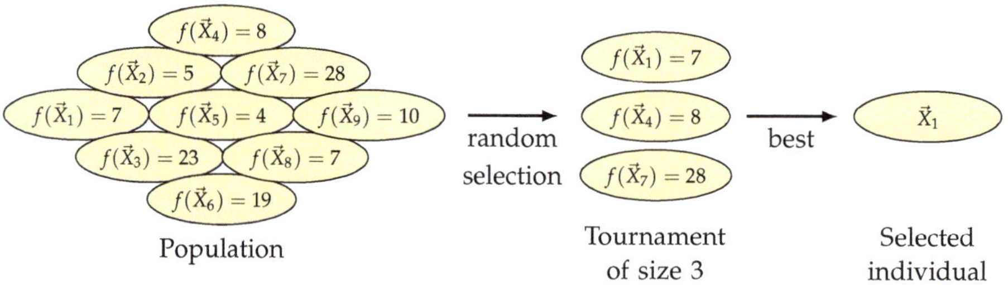

Figure 4.5: Tournament selection with a tournament size of three. Out of the current population, three solutions are chosen randomly. Out of these, the best solution w. r. t. the indicated fitness value is the selected individual.

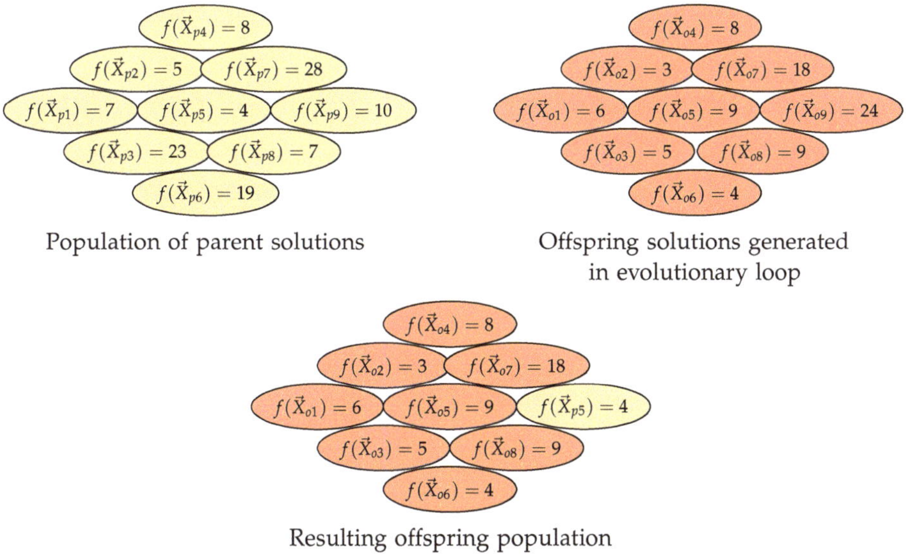

Figure 4.6: Replacement strategy with elitism. The population of parent solutions is depicted in yellow, the offspring solutions generated during the evolutionary loop are depicted in orange. To generate the resulting offspring population, the principle of elitism is applied, and the worst offspring solution is replaced by the best parent solution.

Lemma 4.2 (Guaranees on the solutions generated by the MRTA reoptimization framework with GA)

For any of the MRTA reoptimization problems defined in Chapter 3, i. e. Problems 3.3 to 3.12, the MRTA reoptimization framework with GA in combination with the respective heuristics introduced in Chapter 3 can be applied. This approach ensures the guarantees on the feasibility of the applied heuristic solutions, i. e. Lemmas 3.5, 3.8 and 3.10 to 3.19, to remain valid for the solutions generated by the MRTA reoptimization framework. Furthermore, the upper bounds of the approximation rations for the CMI and TDH, i. e. Theorems 3.2 to 3.5, are preserved.

Proof:

Due to the application of the principle of elitism, the fitness values of the best solution within a population cannot worsen between the evolutionary loops. This ensures the first aspect of Lemma 4.1 to be fulfilled. Since the fitness value is defined such that infeasible solutions always have worse fitness values than feasible solutions (see (4.1)), also the second aspect of Lemma 4.1 is fulfilled, such that Lemma 4.1 applies to the MRTA reoptimization framework with GA as defined in this section. □

The introduced GA-based reoptimization framework combines the heuristic reoptimization approaches of Chapter 3 with a GA suitable for heterogeneous, time-extended MRTA problems with precedence and synchronization constraints. It allows for a broader search of the solution space and thus for a possible improvement of the heuristic solutions. By the application of the principle of elitism together with an appropriate choice of the fitness function, the performance guarantees given for the reoptimization heuristics remain valid for the solutions generated by the GA-based reoptimization framework.

In the following Chapter, a comprehensive evaluation both of the reoptimization heuristics of Chapter 3 and of the GA-based reoptimization framework is given. The proposed reoptimization approaches are furthermore compared to an exact solution approach and to a conventional, randomly initialized, GA.

5 Evaluation and Analysis of the Reoptimization Approaches

An evaluation of the reoptimization heuristics introduced in Chapter 3 and of the GA-based reoptimization framework introduced in Chapter 4 is conducted and the results are presented in this chapter. The evaluation aims at answering the third research question (see Section 2.6) on advantages and disadvantages of the previously introduced reoptimization approaches when applied to heterogeneous, time-extended MRTA reoptimization problem instances and on whether and how their performance is influenced by different problem features. The introduced reoptimization approaches are compared to an exact BnP approach as well as to a conventional, randomly initialized GA. Evaluation criteria are determined to evaluate the performance, responsiveness and stability of the proposed approaches, i. e. the aspects relevant for their application in an interactive MRTA optimization system (see Section 1.1). To evaluate the influence of different features of the MRTA problem instances on the reoptimization approaches of Chapters 3 and 4, three different evaluation scenarios focusing on different features of the MRTA problem instances are introduced. As depicted in Figure 5.1, in evaluation scenario 1, the size of the modified problem instance, i. e. the amount of tasks and agents considered, is varied. In evaluation scenario 2 different levels of heterogeneity w. r. t. different capabilities and velocities of agents as well as varying basic task durations are investigated for a fixed problem size. The influence of different numbers of precedence and synchronization constraints is analyzed in evaluation scenario 3 for a fixed problem size and homogeneous problem instances.

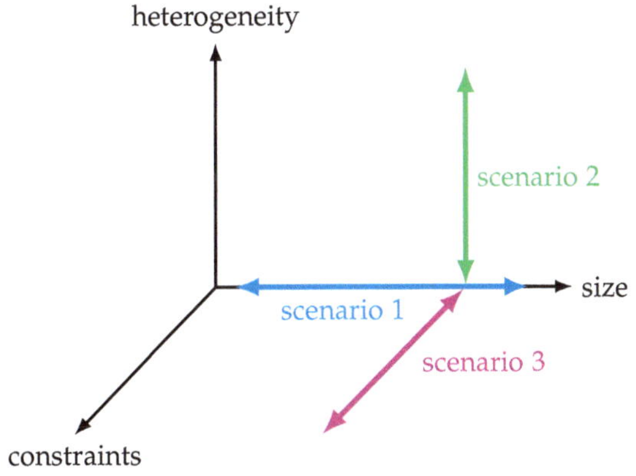

Figure 5.1: Overview of the features varied within each evaluation scenario. More details are given in Table 5.1.

The evaluation setup as well as the evaluation criteria are described in Section 5.1. The results of evaluation scenario 1 investigating the influence of the problem size are presented in Section 5.2. In Section 5.3, the results of evaluation scenario 2 analysing the influence of different sources of heterogeneity are given and in Section 5.4, the results of evaluation scenario 3 varying the number of precedence and synchronization constraints are depicted.

5.1 Evaluation Setup and Evaluation Criteria

In this section, details on the evaluation setup including the parametrization are given. Furthermore, the applied evaluation criteria are introduced.

5.1.1 Common Features of the Evaluation Scenarios

Within each evaluation scenario, different specifications for problem instances are considered. Each specification determines

- the size of the modified MRTA problem instance w. r. t. number of agents and tasks,
- the heterogeneity within the corresponding modified problem instances w. r. t. agents' capabilities and velocities as well as tasks' basic durations,
- and the number of precedence and synchronization constraints to be considered within the modified problem instances.

Within each evaluation scenario, 50 problem instances are determined for each specification and each analyzed modification. Without loss of generality, all distances, velocities and durations within a problem instance are defined dimensionless. Tasks and agents' depots are located on a two-dimensional area of size 100×100. The modified problem instances are solved by the application of

- the respective reoptimization heuristic appropriate for the applied modification as given in Chapter 3,
- the GA-based reoptimization framework specified in Sections 4.2 and 4.3,
- a conventional, randomly initialized GA that applies the same genotype representation, crossover and mutation operators, fitness function, select procedure and replacement strategy with elitism as defined in Section 4.3, i. e. it differs from the GA-based reoptimization framework solely in the initialization, and
- an exact branch-and-price (BnP) solution approach as given in Appendix B.2.

The parameters weighting the task execution times and the waiting times within the objective function (3.1) are set to $\gamma = 1$ and $\epsilon = 0.9$. Thus, transitioning and task execution times are equally weighted, since in several applications they can be expected to be similarly energy consuming. Waiting times are weighted 10% less, since they usually consume less energy, but should still be minimized in order for all tasks to be performed as quickly as possible.

To completely describe the GA-based reoptimization approach as well as the conventional GA, some additional parameters have to be defined. Both approaches apply a constant population size of 50 solutions per generation.[32] As proposed by Entezari and Mahootchi [EM20], the termination criterion is given by a fixed number of iterations, which is set to 100. Since the principle of elitism is applied, which prevents the loss of the best solution between two populations, high crossover and mutation rates allow for a broad search of the solution space. The crossover rate is set to 100% and the mutation rate numbers 50%.

All solution approaches are implemented and run in Python 3.9.6 on an Intel® i5-960K processor with a clock frequency of 3.70 GHz, 16.0 GB RAM, and a Microsoft® Windows 11® operating system.

An overview on the problem specifications investigated in the three evaluation scenarios is given in Table 5.1. In the first evaluation scenario, the problem size is varied, in the second evaluation scenario, different aspects of heterogeneity are investigated, and in the third evaluation scenario, temporal constraints are contained in the problem instances. In the following section, details on evaluation scenario 1 are given.

32 Different population sizes have been tested throughout the bachelor thesis of Teufel [Teu20] which was supervised by the author and a population size of 50 was found to be a good compromise between diversification and computation time needed per evolutionary loop.

Table 5.1: Overview on the problem specifications investigated in the three evaluation scenarios. The aspect varied within the problem specifications of an evaluation scenario is printed in italics.

	Problem size	Heterogeneity	Temporal constraints
Evaluation scenario 1	$1-10$ *tasks and* $1-10$ *agents*	Homogeneous problem instances	None
Evaluation scenario 2	3 agents and 8 tasks	*different agent capability, agent velocity and task duration levels*	None
Evaluation scenario 3	3 agents and 8 tasks	Homogeneous problem instances	$0-2$ *precedence and synchronization constraints*

5.1.2 Specification of Evaluation Scenario 1 – Different Sizes of Problem Instances

In the first evaluation scenario, the influence of the number of tasks and agents within the modified problem instance is analyzed. The goal is furthermore to determine a meaningful size for the problem instances of the following evaluation scenarios. The problem instances are all homogeneous, i. e. all basic task durations are set to ten, i. e. $\tau_i = 10 \, \forall i \in \mathcal{N}$, all agents $m \in \mathcal{K}$ have a capability of $a_i^m = 1$ for all tasks $i \in \mathcal{N}$, and the velocities of all agents $m \in \mathcal{K}$ is set to $v_m = 1$. No precedence or synchronization constraints are considered in the problem instances.

Within this evaluation scenario, the modifications of task insertion and task deletion are analyzed since they are the most common. Furthermore, the solutions generated by the respective reoptimization approaches are not only guaranteed to be feasible, moreover upper bounds on the resulting approximation ratios can be given (see Sections 3.2.5 and 3.3.5). Given the initial problem instances, the modified problem instances are established as follows: To determine the modified problem instances for the task insertion, an additional task, whose position is determined randomly out of a uniform distribution over the whole area, is added to the initial problem instance. For the modification of task deletion, the task to be deleted is selected randomly out of the existing tasks within the initial problem instance.

The evaluation scenario has two parts: In the first part, the number of agents is fixed and the number of tasks is varied, while in the second part it is vice versa. To determine the initial problem instances for the first part of the evaluation scenario, the number of agents is set to three, which is a common number for evaluations in heterogeneous, time-extended MRTA problems [Kor11, KKB+12]. The number of tasks within the initial problem instance is increased stepwise. For each number of tasks, 50 problem instances are generated. For the first 50 initial problem instances containing one task, the position of the tasks and the agents' depots is chosen randomly out of a uniform distribution over the whole area. In order to minimize the influence of other aspects like the tasks' and depots' positions on the evaluation results, the problem instances are defined iteratively: For the initial 50 initial problem instances containing two tasks, the positions of the agents depots' and the first task is equivalent to their corresponding problem instance containing one task. The position of the second task is again chosen randomly out of a uniform distribution. The initial problem instances containing three tasks are defined based on the problem instances containing two tasks and so forth. Overall, initial problem instance containing one to ten tasks are generated.

In the second part of the evaluation scenario, the number of agents within the problem instances is varied from one to ten agents. The fixed number of tasks contained in each modified problem instance is set to eight, i. e. the initial problem instances generated for the modification of task insertion contain seven tasks and the ones for the modification of task deletion contain nine tasks. To reduce the influence of the tasks' positions, they are determined once for the 50 problem instances for each modification and remain fixed, independently of the number of agents considered. The agents' depot positions are determined iteratively as described for the tasks' positions in the first part of the evaluation scenario.

The results of the first evaluation scenario build the basis for defining a problem size that yields a good compromise between complexity and calculability for the second evaluation scenario, which is described in the following section.

5.1.3 Specification of Evaluation Scenario 2 – Different Heterogeneity Levels

In this evaluation scenario, the influence of different sources of heterogeneity is analyzed. Therefore, different levels of agents' capabilities, agents' velocities and basic task durations are defined. The problem instances of this evaluation scenario do neither contain precedence nor synchronization constraints. All modifications unrelated to precedence and/or synchronization constraints (i. e. the modifications of task insertion, task deletion, task position variation, task duration variation, agent capability variation and agent velocity variation) are considered. Based on the results of the previous evaluation scenario 1, each modified problem instance consists of three agents and eight tasks.

To represent different possible strengths of the three agents w. r. t. their capabilities to perform different tasks, tasks are assumed to be of three different types and each task is assigned to one of these three task types. Using these task types, the different capability levels are defined as given in Table 5.2. Capability level 0 corresponds to the case of homogeneous capabilities, i. e. every agent is fully capable to accomplish each task, independent of its type. With capability level 1, each agent is still capable of accomplishing each task type, but for each task type there are agents that take up to double the time to finish the corresponding tasks compared to other agents. In capability level 2, tasks of type 1 and 3 can only be performed by two of the three agents and the differences in speed of operation remain similar as in level 1. The agent-task-assignment for tasks of type 1 and 3 becomes unambiguous for capability level 3, since these tasks can only be performed by one of the three agents. For tasks of type 2, still all agents possess some capability to different amounts. Since the case of an unambiguous assignment of all tasks, i. e. for each task only one agent being capable of its execution, is trivial and reduces the time-extended MRTA problem to a pure scheduling problem, it is not considered.

Similarly to the capability levels, also different levels of the agents' velocities are defined, which are depicted in Table 5.3. Also for the velocities, level 0 corresponds to the homogeneous case where there are no differences between the agents' velocities. In levels 1 and 2, the differences in the agents' velocities increase with agent k_1 being the slowest and agent k_3 the fastest one.

For the basic task durations, different levels are defined as well. They are given in Table 5.4. Analogous to the capability and velocity levels, duration level 0 corresponds

Table 5.2: Definition of different capability levels.

Capability level	Agent	Task type 1	Task type 2	Task type 3
capability level 0	k_1	1	1	1
	k_2	1	1	1
	k_3	1	1	1
capability level 1	k_1	1	0.8	0.5
	k_2	0.8	0.5	1
	k_3	0.5	1	0.8
capability level 2	k_1	1	0.8	0
	k_2	0.8	0.5	1
	k_3	0	1	0.5
capability level 3	k_1	1	0.8	0
	k_2	0	0.5	1
	k_3	0	1	0

Table 5.3: Definition of different velocity levels.

Velocity level	Agent k_1	Agent k_2	Agent k_3
velocity level 0	1	1	1
velocity level 1	0.8	0.9	1
velocity level 2	0.4	0.7	1

to the homogeneous case with every task of each task type having the basic duration of $\tau_i = 10, \forall i \in \mathcal{N}$. The differences in the task types' basic durations increase for duration levels 1 and 2, with tasks of type 1 always requiring the least and tasks of type 3 requiring the most processing time.

To generate the initial problem instances for this evaluation scenario, homogeneous instances with capability, velocity and duration levels of 0 are generated. The homogeneous initial problem instances build the basis for the initial problem instances with different capability, velocity or duration levels. Independent of the capability, velocity and duration level under consideration, the task types as well as the positions of the tasks and the agents' depots remain as defined in the homogeneous initial problem instances. Reason for this is to prevent influences of altered task types or position within the results. In order for all modified problem instances to contain eight tasks as previously specified, for the modification of task insertion, 50 homogeneous problem instances with seven tasks are generated, for the modification of task deletion, 50 homogeneous instances with nine tasks and for all other modifications, 50 homogeneous problem instances with eight tasks are generated. For these instances, the positions of the tasks and the agents' depots are chosen randomly from a uniform distribution over the whole area. Also, for each task, the task type is randomly set to be either 1, 2 or 3.

Independently of the heterogeneity, velocity and duration level under consideration, the modifications are chosen identically for all corresponding problem instances. For the task insertion, the new task is randomly assigned to one task type and its position is chosen randomly out of the uniform distribution over the whole area. For all 50 problem instances the task's position and type remain fixed, independently of the heterogeneity, velocity and duration level under consideration. In the same manner, for all 50 problem instances the task to be deleted for the modification of task deletion, is initially chosen randomly out of all tasks and remains fixed for all heterogeneity levels. For the modification of task position variation, one task within the initial problem instance is chosen randomly. Within a radius of 50 of its initial position, a new position is chosen by chance. It is ensured, that the new position lies within the area of size 100×100, which is generally defined for all problem instances (see Section 5.1.1). The modified task and its new position are employed within every heterogeneity level. Also for the task duration variation, for all 50 problem instances one task to be modified is chosen randomly. Its duration is varied by a factor chosen uniformly out of

Table 5.4: Definition of different levels of basic task durations.

Duration level	Task type 1	Task type 2	Task type 3
duration level 0	10	10	10
duration level 1	5	10	15
duration level 2	1	10	50

the set $[0.5, 0.9] \cup [1.1, 2.0]$, for a shorter or longer task duration, respectively. The same factor is applied to the same task independently of the heterogeneity level. An analogous approach is applied for the modifications of agent capability variation and agent velocity variation. The agent to be modified, which is chosen randomly, is the same for all heterogeneity levels. For the modification of capability variation, one task type is determined randomly and the agent's capability for this task type is varied by a random factor out of the set $[0.5, 0.9] \cup [1.1, 2.0]$. In the same manner, the factor for modifying the agent's velocity is chosen randomly out of the set $[0.5, 0.9] \cup [1.1, 2.0]$ for the modification of agent velocity variation.

Within the second evaluation scenario, precedence and synchronization constraints contained within the problem instances are not considered. Their influence is investigated in the third evaluation scenario, which is described in the following section.

5.1.4 Specification of Evaluation Scenario 3 – Temporal Constraints

In the third evaluation scenario, the influence of precedence and synchronization constraints on the results of the reoptimization approaches shall be evaluated. As in evaluation scenario 2, all modified problem instances contain three agents and eight tasks. Furthermore, only homogeneous problem instances are considered, i. e. all instances are of capability, velocity and duration level 0 (see Section 5.1.3).

Upon this, six MRTA problem specifications differing in the number of precedence and synchronization constraints are determined. The homogeneous initial problem instances without any precedence or synchronization constraints of evaluation scenario 2 (see Section 5.1.3) serve as the basis problem specification. The other five specifications contain one precedence constraint, two precedence constraints, one synchronization constraint, two synchronization constraints, or one precedence and one synchronization constraint, as given in Table 5.5. To determine the corresponding initial problem instances, the respective number and type of constraints is added to the unconstrained basic problem instances. The corresponding constraints always consider two tasks, which are chosen randomly out of all tasks. For precedence constraints also the order

Table 5.5: Number of synchronization and precedence constraints contained within the different problem specifications of evaluation scenario 3.

Problem specification	# synchronization constraints	# precedence constraints
sync. 0 prec. 0	0	0
sync. 0 prec. 1	0	1
sync. 0 prec. 2	0	2
sync. 1 prec. 0	1	0
sync. 2 prec. 0	2	0
sync. 1 prec. 1	1	1

of the tasks, i. e. which task is set to be the precedent one, is chosen randomly. If two synchronization constraints are added and both of them contain one common task, e. g. $s_{1,3} = 1$ and $s_{1,6} = 1$, this corresponds to all three tasks having to be performed synchronously, i. e. tasks n_1, n_3 and n_6 in the example. Since non-empty solution spaces must be ensured, all initial instances generated are checked for feasible definitions of the temporal constraints. If the randomly chosen constraints contradict each other, e. g. $p_{1,3} = 1$ and $p_{3,1} = 1$, new constraints are chosen randomly. To prevent the influence of task or depot positions on the results, the positions of the tasks and the agents' depots remain unchanged w. r. t. the unconstrained initial problem instances.

Within this evaluation scenario, all modifications analyzed in Chapter 3 are evaluated. The generation of the modified problem instances for the modifications of task insertion, task deletion, task position variation, task duration variation, agent capability variation and agent velocity variation, is equivalent as described for evaluation scenario 2 (see Section 5.1.3). For the modifications of precedence or synchronization constraint deletion, one of the existing constraints of the respective type is randomly chosen to be deleted. Obviously, this modification is only applied to the problem instances with at least one of the respective constraints. For the modifications of precedence or synchronization constraint insertion, a respective constraint for two of the existing tasks is added to the problem instance in the same random manner, as described above for the initial problem instances. Furthermore, it is ensured that the resulting modified problem instances can be solved by the application of the eDIH (see Section 3.8.2). If the additional constraint leads to an unsolvable problem instance, it is replaced by a newly generated constraint.

So far, the evaluation scenario considered for the analysis have been introduced in detail in this chapter. The following section determines the criteria used to evaluate the reoptimization results.

5.1.5 Evaluation Criteria

As outlined in Section 1.1, in the context of interactive MRTA optimization systems, three aspects are relevant when solving the modified problem instances. *High quality solutions* w. r. t. the objective function (3.1) are relevant for the MRS performance [GM04, VR21]. Furthermore, user acceptance is heavily influenced by the *responsiveness* of the solution method applied to the modified problem instance as well as by the *stability* of the solution [HGQ+12, MKF+15]. To evaluate these aspects, the following measures are applied:

To assess the quality of approximative optimization or reoptimization approaches, the **approximation ratio** α as given by Definition 2.11 is an appropriate measure (see Section 2.1). Hence, for the solutions generated by the reoptimization heuristics, the GA-based reoptimization framework and the conventional, randomly initialized GA, the approximation ratios w. r. t. the objective function for heterogeneous, time-extended MRTA problems with precedence and synchronization constraints as given by (3.1) are determined to evaluate their quality.[33] The theoretical guarantees on the approximation ratio derived for the heuristics in this thesis and for similar TSP reoptimization problems are at most two (see Sections 3.2.5 and 3.3.5 as well as Table 2.4). However, the guaranteed approximation ratios are strictly conservative. Hence, it is expected that the approximation ratio for the majority of problem instances in practical application is much smaller. In the TSP, the smallest guaranteed upper bound on the approximation ratio for reoptimization approaches equals $4/3$ (see Table 2.4). Taking some margin from this value, the following evaluation considers approximation ratios $\alpha \leq 1.2$ as desirable.[34]

To evaluate the responsiveness of the applied solution methods, their **calculation time** needed to solve the modified problem instance is the appropriate measure (see Section 1.1). It is assessed for each modified problem instance for all four solution approaches, i. e. the reoptimization heuristics, the GA-based reoptimization framework, the conventional, randomly initialized GA and the BnP. Different classifications of acceptable system response time categories that depend for example on users' expectations [Mil68, Seo08] or on task complexity [Shn87] can be found in literature. Doherty and Sorenson [DS15] combine and expand these frameworks focusing on perceived user experience. In general, system response times below 300 ms are perceived as instantaneous, closed-loop interactions [DS15]. Response delays of up to 1 s can be detected by the human, but are still perceived as immediate interaction [DS15]. User might even accept longer response times up to 10 s and stay focused on the interaction if they receive appropriate feedback during the delay [DS15]. Longer system response times however are likely to cause the user to multitask during the process or even to disengage with the task if system response times are above 5 min.

To evaluate the stability of the solutions generated for the modified problem instance, a measure for their difference to the initial solution must be defined (see Section 1.1). As introduced in Section 3.1.1, a solution to a time-extended MRTA problem instance contains the routing information together with the timing information, i. e. $\vec{X}_{\mathcal{I}} = X_{\mathcal{I}} \cup \{t_i | i \in V\}$. However, for the solution approaches under consideration in this thesis, a measure of the difference between the routings of the initial solution and the solution to the modified problem instance fully describes the solution differences. This is due to the fact, that, given a certain routing, all reoptimization heuristics (see Algorithms Algorithm 1, 2, 3, 5 and 6), as well as the GA applied within the GA-based reoptimization framework and within the conventional GA (see genotype representation in Section 4.3) and, by definition, also the exact BnP approach determine the corresponding optimal timing information according to (3.5). Thus, the routing generated by any of the applied solution approaches inherently also determines the timing of the corresponding solution.

[33] By definition, the exact solution generated by the BnP approach has an approximation ratio of $\alpha = 1$, which is why this is not given explicitly. The BnP solution is needed to determine the approximation ratios of the solutions generated by the other solution approaches.

[34] To the best of the authors knowledge, so far no commonly acknowledged threshold for approximative solution approaches in MRTA applications exists. Therefore, this choice has been made for the evaluation within this thesis. However, further verification in various application fields is necessary.

To measure the difference between the initial routing and the routing for the modified problem instance, and thus of the corresponding solutions, the **Levenshtein distance** is applied. The Levenshtein distance, named after the Russian mathematician Vladimir Levenshtein, is one of the most popular measures in string matching [Lis13, Chapter 8.2]. It is given by the minimum number of basic operations required to convert one string into another. The considered basic operations include the insertion of a character, the deletion of a character, and the replacement of a character by another [Lis13, Chapter 8.2]. Interpreting the routes of the agents as strings with the vertices representing the characters, the Levenshtein distance between two routes is given as the minimum number of operations including

- the insertion of a vertex
- the deletion of a vertex
- the replacement of a vertex by another

required to convert one route into another.[35] In order to evaluate the difference between the initial routing $X_{\mathcal{I}}$ and the routing $X_{\mathcal{I}_{\mathrm{mod}}}$ generated by the solution approach for the modified problem instance, the sum over the Levenshtein distances of all agents' routings is determined. Denoting the Levenshtein distance as lev, the Levenshtein distance between the routings $X_{\mathcal{I}}$ and $X_{\mathcal{I}_{\mathrm{mod}}}$ is given by

$$\mathrm{lev}(X_{\mathcal{I}}, X_{\mathcal{I}_{\mathrm{mod}}}) = \sum_{m \in \mathcal{K}} \mathrm{lev}(x_{\mathcal{I}}^{m}, x_{\mathcal{I}_{\mathrm{mod}}}^{m}), \tag{5.1}$$

with $x_{\mathcal{I}}^{m}, x_{\mathcal{I}_{\mathrm{mod}}}^{m}$ denoting the route of agent $m \in \mathcal{K}$ in the solutions to the initial instance $\mathcal{I}$ and the modified instance $\mathcal{I}_{\mathrm{mod}}$, respectively. To assess the Levenshtein distances in the context of time-extended MRTA reoptimization problems, the analogy to string matching is adduced. According to Lisbach and Meyer [Lis13, Chapter 8.2], acceptable Levenshtein distances depend on the size of the problem. For problems up to the size considered in the following evaluation scenarios, the acceptable Levenshtein distances range from zero to two.

As outlined in the previous sections, within each evaluation scenario, for every evaluation specification 50 initial and modified MRTA problem instances are generated. They share the same number of agents and tasks, the heterogeneity levels and the number of precedence and synchronization constraints (see Section 5.1.1). For each problem instance and every solution approach applied to it, the approximation ratio, calculation time and Levenshtein distance are determined. In the following presentation of the results, the corresponding average values as well as the standard deviations are given.

For some modifications in some evaluation scenario, additionally the convergence of the GA-based reoptimization framework and the conventional, randomly initialized GA is compared. Since the principle of elitism is applied, the approximation ratio of

[35] For example, the Levenshtein distance between the routes $\{o_1, n_2, n_3, n_5, o_1\}$ and $\{o_1, n_3, n_5, o_1\}$ equals one.

the best solution contained within the populations cannot deteriorate in the course of the iterations. Due to the fixed number of 100 iterations conducted by both GAs, the convergence behavior is not explicitly contained in the computation time or the other measures. To gain some insight into the convergence behavior, it is evaluated for how many of the 50 problem instances, the respective GA reaches certain thresholds of the approximation ratio, typically $\alpha \leq 1.5$, $\alpha \leq 1.2$ and $\alpha \leq 1.1$. Out of these problem instances, that converged to a certain approximation ratio threshold, additionally the average number of iterations that were required to reach that threshold, is given.

Having introduced the evaluation scenarios and the evaluation criteria applied to analyze the reoptimization approaches presented in Chapters 3 and 4, the results of the first evaluation scenario, varying the size of the problem instances, are given in the following section.

5.2 Evaluation of Different Sizes of Problem Instances

In this section, the results of the first evaluation scenario are presented. As introduced in Section 5.1.2, homogeneous problem instances without any precedence or synchronization constraints are considered and the influence of the number of tasks and agents within the modified problem instances on the results generated by the different solution approaches is investigated. A presentation of the results is given in the following section, followed by an analysis thereof in Section 5.2.2.

5.2.1 Evaluation Results

A presentation of the results of the evaluation of the influence of the number of tasks and agents within the modified problem instances on the solutions generated for the modifications of task insertion and task deletion is given in this section. First, the results on the approximation ratios are presented, followed by the calculation times and the Levenshtein distances.

Approximation ratio

For the modification of task insertion, the approximation ratio of the solutions generated by the CMI, the GA-based reoptimization framework and the conventional, randomly initialized GA for modified MRTA problem instances with three agents and different numbers of tasks are depicted by a box plot given in Figure 5.2. The small problem instances with up to three tasks are all solved to optimality by both GA-based solution approaches. By the application of the CMI, 7 out of 50 problem instances are not solved to optimality for both problem sizes with the worst resulting approximation ratio being 1.226. For the problem instances containing four to ten tasks, the

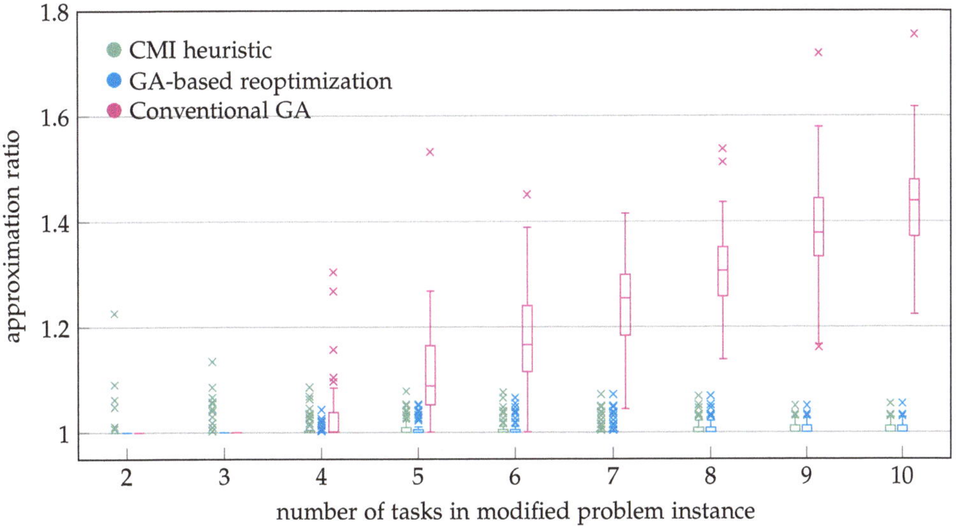

Figure 5.2: Evaluation scenario 1: Approximation ratios for the modification of task insertion over problem instances with different numbers of tasks. Detailed values are given in Table C.1 in Appendix C.1.

approximation ratios generated by the CMI remain relatively stable for different numbers of tasks contained in the modified problem instances. The average approximation ratio varies between 1.011 and 1.007 with a slight tendency towards the smaller approximation ratios for modified problem instances with more tasks (see Table C.1 in Appendix C.1). For the problem instances containing up to six tasks, the GA-based reoptimization framework yields on average slightly smaller approximation ratios than the CMI. For the modified problem instances containing seven to ten tasks, the approximation ratios of the CMI and the GA-based reoptimization framework are identical. In contrast to this, the approximation ratios of the conventional GA show an approximately linear increase for the problem instances with four to ten tasks, yielding clearly higher approximation ratios than the two reoptimization approaches.

The respective results on the approximation ratios generated by the CMI for modified problem instances with eight tasks and different numbers of agents are depicted in Figure 5.3. Independently of the number of agents considered within the problem instances, the approximation ratios of the results generated by the TDH and the GA-based reoptimization framework are identical. For these reoptimization approaches, the resulting approximation ratios are equal to 1.0 for the majority of problem instances, with the highest approximation ratio being equal to 1.083 for a modified problem instance with two agents. With an increase in the number of agents, the spread of the approximation ratios slightly decreases. Contrary to the reoptimization approaches, the conventional GA shows an increase in the approximation ratios with a growing number of agents contained in the modified problem instance. The incline is approximately linear w. r. t. the median values for problem instances with one to ten agents. Independently of the number of agents contained in the modified problem in-

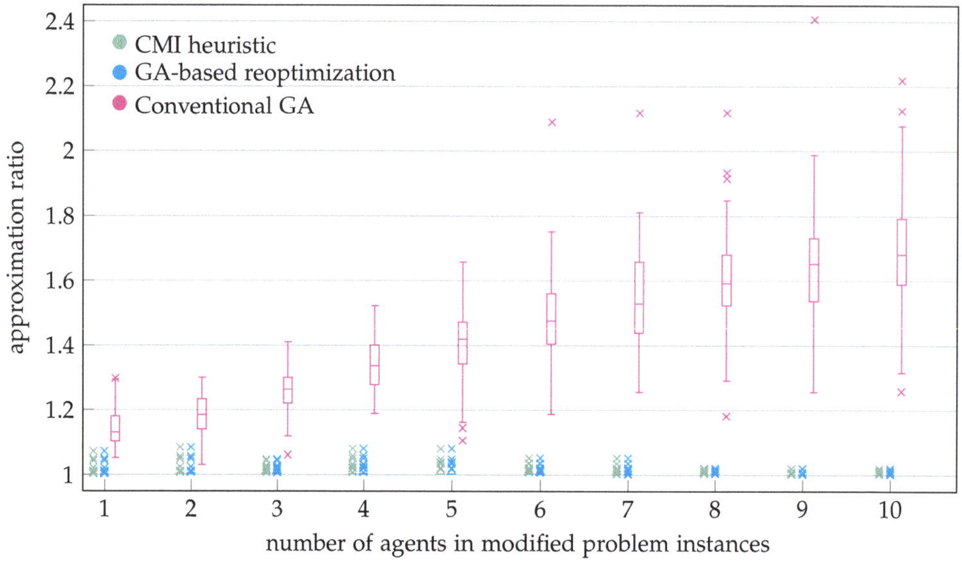

Figure 5.3: Evaluation scenario 1: Approximation ratios for the modification of task insertion over problem instances with different numbers of agents. Detailed values are given in Table C.1 in Appendix C.1.

stances, the average approximation ratio of the conventional GA is considerably higher than the one of the reoptimization approaches, and no problem instance is solved to optimality by the conventional GA. For individual problem instances containing six and more agents, the approximation ratio of the solution generated by the GA exceeds a value of $\alpha = 2$.

The results on the approximation ratio for the modification of task deletion show similar dependencies on the numbers of tasks and agents contained in the modified problem instances. The mean values as well as the corresponding standard deviations for the specifications investigated are given in Table 5.6.[36] For the modified problem instances with up to three agents, both GA approaches solve all instances to optimality. By the application of the TDH to these problem instances with up to three tasks, between three and ten out of the 50 problem instances are not solved to optimality. For the problem instances with one task, the highest approximation ratio equals 1.28, with two tasks it is equal to 1.06, and with three tasks the highest approximation ratio is given by 1.10. On average, the approximation ratios generated by the reoptimization approaches are close to 1.0 and have a standard deviation smaller than 0.02 for the problem instances with more than one task. For the modified problem instances having six or more tasks, the approximation ratio of the results generated by the TDH and the GA-based reoptimization framework are identical. The average approximation ratios resulting from the conventional, randomly initialized GA, increase approximately

[36] As the qualitative results for the modification of task insertion and task deletion are similar, the results for task deletion are given in a table only.

Table 5.6: Evaluation scenario 1: Results on the approximation ratio α (mean and standard deviation (SD)) the for the modification of task deletion.

Variation			TDH heuristic		GA-based reoptimization		Conventional GA	
			mean	SD	mean	SD	mean	SD
Task deletion	Number of tasks	1	1.0134	0.0538	1.0	0.0	1.0	0.0
		2	1.0021	0.0091	1.0	0.0	1.0	0.0
		3	1.0049	0.0165	1.0	0.0	1.0	0.0
		4	1.0051	0.0161	1.0008	0.0038	1.0242	0.0371
		5	1.0047	0.0123	1.0026	0.008	1.0825	0.0738
		6	1.0042	0.012	1.0042	0.012	1.1584	0.0695
		7	1.0047	0.0133	1.0047	0.0133	1.2469	0.0904
		8	1.0032	0.0094	1.0032	0.0094	1.3126	0.0996
		9	1.0038	0.0102	1.0038	0.0102	1.373	0.0874
	Number of agents	1	1.0023	0.0061	1.0023	0.0061	1.1281	0.0462
		2	1.002	0.0076	1.002	0.0076	1.1763	0.0576
		3	1.0017	0.0047	1.0017	0.0047	1.2584	0.0839
		4	1.0029	0.006	1.0029	0.006	1.3406	0.098
		5	1.0038	0.0077	1.0038	0.0077	1.4081	0.1349
		6	1.0048	0.01	1.0048	0.01	1.4605	0.1395
		7	1.0084	0.0188	1.0084	0.0188	1.5196	0.1388
		8	1.0082	0.0196	1.0082	0.0196	1.5605	0.1425
		9	1.005	0.0111	1.005	0.0111	1.626	0.1704
		10	1.0037	0.0095	1.0037	0.0095	1.6663	0.1595

linearly with the number of tasks contained in the modified problem instances for instances with more than three tasks to an average of $\alpha = 1.37$ for modified instances with nine tasks. Compared to the GA-based reoptimization framework, the standard deviation of the conventional GA is about six to ten times higher for problem instances with at least four tasks. Independently of the number of agents, the results for the TDH and the GA-based reoptimization framework are identical. The average approximation ratios are close to 1.0, varying between about 1.002 and 1.008 for different numbers of agents. For the conventional GA however, the average approximation ratios are higher for all numbers of agents investigated and increase with the number of agents from 1.128 with one agent to 1.666 with ten agents. The standard deviations are about seven to 17 times higher than the ones of the TDH and the GA-based reoptimization framework.

The differences between the GA-based reoptimization and the conventional GA also becomes obvious, when comparing the convergence behavior of the two GA-based approaches. Table 5.7 gives an overview of how many of the 50 problem instances reach an approximation ratio threshold of $\alpha \leq 1.1$ for the GA-based reoptimization and how many iterations it takes for these instances on average to converge to this threshold. The same information is given for the conventional GA for reaching thresholds of

$\alpha \leq 1.5$, $\alpha \leq 1.2$ and $\alpha \leq 1.1$. Independent of the numbers of tasks and agents contained in the modified problem instances, both GA-based reoptimization frameworks for the modifications of task insertion and task deletion reach an approximation ratio threshold of $\alpha \leq 1.1$ for all 50 problem instances, which in the vast majority of problem specifications takes on average exactly one iteration to be reached. In contrast to this, the conventional, randomly initialized GA reaches an approximation ratio threshold of $\alpha \leq 1.1$ for all 50 problem instances only for the specifications with up to three tasks for both modifications. The more tasks are contained in the modified problem instances, the less often the GA converges to a threshold of $\alpha \leq 1.1$, which then also takes more iterations. For example, for the problem instances containing six tasks, in 10 out of 50 problem instances the conventional GA reaches an approximation ratio threshold of $\alpha \leq 1.1$ for both modifications, which are reached on average after 18.6 and 13.5 iterations, respectively. For the specifications with a growing number of agents, the GA does not converge to an approximation ratio threshold of $\alpha \leq 1.1$ for any problem instance with more than three agents, except for one problem instance with five agents for the modification of task deletion. While for the specifications with growing numbers of tasks, only within the problem instances with eight and more tasks some do not converge to a threshold of $\alpha \leq 1.5$, less than half of the problem instances with seven or more agents converge to an approximation ratio threshold of $\alpha \leq 1.5$ for both modifications by the application of the conventional GA.

Calculation time

In Figure 5.4, a box plot of the calculation times required by the different solution approaches to solve the task insertion problem instances for different numbers of tasks contained in the modified problem instance is depicted. For a better ascertainability it is cut off at a maximum of 90 s. The average calculation times required by the CMI remain below 3 ms for all evaluated numbers of tasks contained in the modified problem instance (see average values given in Table C.2 in Appendix C.1). For both GA-based solution approaches, the required calculation times are very similar. They grow approximately linear with the number of tasks contained in the problem instance and remain below 25 s for all numbers of tasks contained in the problem instances. Contrary to this, the calculation time required by the exact BnP approach and also its spread grow exponentially and surpasses the GA approaches for the instances containing nine and ten tasks. For problem instances with ten tasks, the average BnP calculation time is 373 s (see Table C.2 in Appendix C.1) and the maximum calculation time required by one of the problem instances equals 1055 s.

The box plot of the calculation times required for different numbers of agents contained within the task insertion problem instance is depicted in Figure 5.5. Contrary to different numbers of tasks, the calculation time required by the BnP approach grows approximately linear with the number of agents contained within the problem instance. On average, the BnP approach requires less computation time than the GA-based reoptimization framework and the conventional GA for all numbers of agents

Table 5.7: Evaluation scenario 1: Convergence of the GA-based reoptimization framework and the conventional, randomly initialized GA for the modifications of task insertion and task deletion.

		Variation	GA-based reoptimization iterations to $\alpha \leq 1.1$		Conventional GA iterations to $\alpha \leq 1.5$		Conventional GA iterations to $\alpha \leq 1.2$		Conventional GA iterations to $\alpha \leq 1.1$	
			mean	# inst.	mean	# inst.	mean	# inst.	mean	# inst.
Task insertion	Number of tasks	2	1.00	50	1.00	50	1.00	50	1.00	50
		3	1.18	50	1.00	50	1.14	50	2.02	50
		4	1.00	50	2.16	50	4.33	48	13.39	46
		5	1.00	50	1.00	49	9.80	44	12.64	28
		6	1.00	50	1.08	50	11.48	29	18.60	10
		7	1.00	50	1.16	50	23.93	15	1.00	1
		8	1.00	50	1.75	48	9.75	4	–	0
		9	1.00	50	8.26	46	22.00	4	–	0
		10	1.00	50	12.80	41	–	0	–	0
	Number of agents	1	1.00	50	1.00	50	3.71	41	3.55	11
		2	1.00	50	1.00	50	15.75	28	12.75	4
		3	1.00	50	1.14	50	25.89	9	73.00	1
		4	1.00	50	1.72	47	1.00	1	–	0
		5	1.00	50	5.44	41	1.50	4	–	0
		6	1.00	50	1.97	29	36.00	1	–	0
		7	1.00	50	2.71	21	–	0	–	0
		8	1.00	50	6.17	12	1.00	1	–	0
		9	1.00	50	1.11	9	–	0	–	0
		10	1.00	50	1.86	7	–	0	–	0
Task deletion	Number of tasks	1	1.00	50	1.00	50	1.00	50	1.00	50
		2	1.00	50	1.00	50	1.00	50	1.00	50
		3	1.00	50	1.00	50	1.02	50	1.22	50
		4	1.00	50	1.00	50	1.64	50	8.90	48
		5	1.00	50	1.00	50	8.15	47	16.53	30
		6	1.00	50	1.00	50	16.06	36	13.50	10
		7	1.00	50	1.58	50	17.08	13	1.33	3
		8	1.00	50	5.78	46	14.17	6	65.00	1
		9	1.00	50	12.15	47	1.00	2	–	0
	Number of agents	1	1.00	50	1.00	50	4.20	46	4.69	13
		2	1.00	50	1.00	50	11.06	31	17.29	7
		3	1.00	50	1.36	50	17.40	15	–	0
		4	1.00	50	1.69	45	10.67	3	–	0
		5	1.00	50	3.53	38	1.00	2	1.00	1
		6	1.00	50	6.19	32	1.00	1	–	0
		7	1.00	50	7.22	23	–	0	–	0
		8	1.00	50	8.27	22	–	0	–	0
		9	1.00	50	13.55	11	–	0	–	0
		10	1.00	50	1.00	5	–	0	–	0

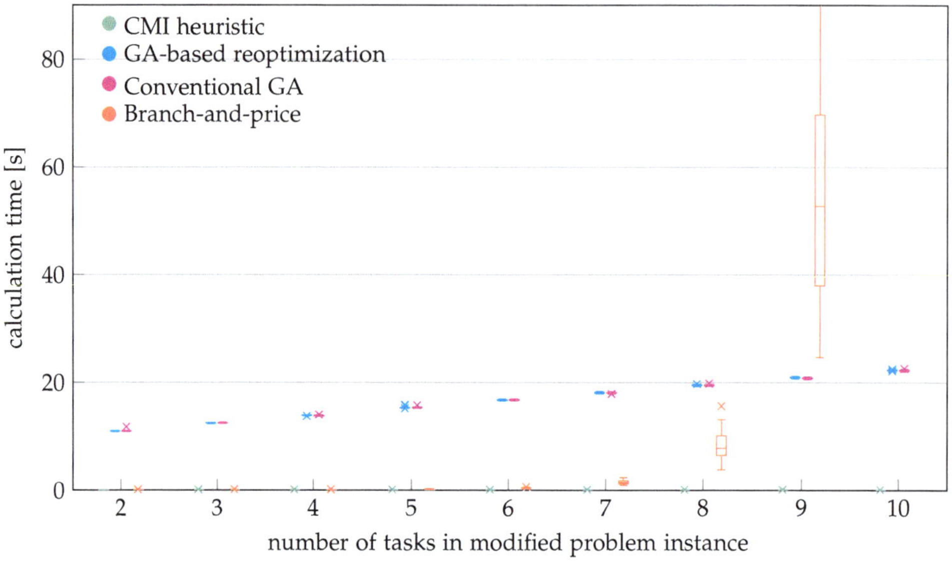

Figure 5.4: Evaluation scenario 1: Calculation times for the modification of task insertion over problem instances with different numbers of tasks. Detailed values are given in Table C.2 in Appendix C.1.

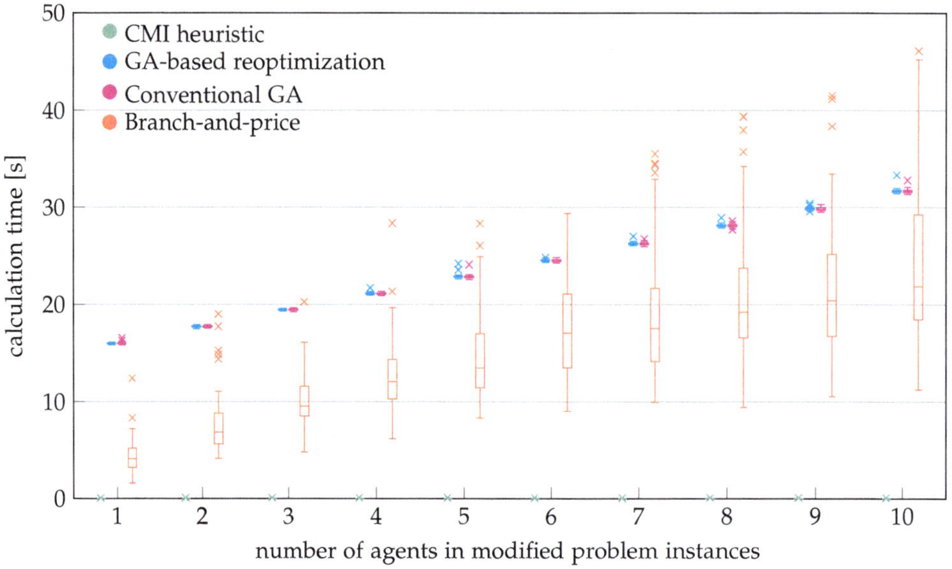

Figure 5.5: Evaluation scenario 1: Calculation times for the modification of task insertion over problem instances with different numbers of agents. Detailed values are given in Table C.2 in Appendix C.1.

evaluated. The BnP calculation time clearly has the greatest spread of all solution approaches. The calculation times required by the GA-based reoptimization framework and the conventional GA are very similar, and they grow approximately linear with the number of agents. The by far smallest calculation time with average values below 4 ms is required by the CMI heuristic.

An overview of the calculation times required by the different approaches to solve the task deletion problem instances is given in Table 5.8. In the table, the mean value as well as the standard deviation over the 50 problem instances contained in each problem specification is given. The TDH requires on average a calculation time of less than 4 ms to solve all problem instances. The required calculation time grows slightly with the number of tasks and with the number of agents contained in the modified problem instances. The standard deviations are within a range of up to 6 ms. As for the modification of task insertion, for any problem specification evaluated, i. e. number of tasks and agents, the average calculation times required by the GA-based reoptimization framework and by the conventional, randomly initialized GA, are very similar. Both grow approximately linear with the number of tasks and with the number of agents contained in the modified problem instance. They are roughly 10^4 times higher than the calculation times required by the TDH reoptimization heuristic and have a standard deviation of less than 0.3 s for all numbers of agents and tasks evaluated. The results on the calculation time required by the BnP approach for the task deletion problem instances match the results for the task insertion reoptimization approaches. The calculation time grows exponentially with the numbers of tasks contained within the problem instances and surpasses the calculation times required by the GA approaches for instances with nine and more tasks. Also, the standard deviation of the calculation time grows exponentially with the number of tasks contained in the problem instances. Contrary to this, the average computation times and the standard deviations are influenced approximately linearly by the number of agents within the problem instance.

Levenshtein distance

For the modification of task insertion, the Levenshtein distances of the initial solution to the solutions generated by the CMI, the GA-based reoptimization framework, the conventional GA and the exact BnP approach are depicted over the varying number of tasks in Figure 5.6. By definition, the routings of the solutions generated by the CMI (see Algorithm 1) differ by one additional task contained in the route of one agent. Thus, the Levenshtein distance of the CMI is always equal to 1, independently of the number of tasks or agents contained within the modified problem instances. For the problem specifications with more than five tasks, also the Levenshtein distances of the solutions generated by the GA-based reoptimization framework are equal to 1 for all 50 problem instances. For problem instances with fewer tasks, there exist some outliers having a Levenshtein distances up to 8 for individual problem instances. The Levenshtein distances of the solutions generated by the conventional GA and the exact

Table 5.8: Evaluation scenario 1: Results on the calculation time in seconds (mean and standard deviation (SD)) the for the modification of task deletion.

		TDH heuristic		GA-based reoptimization		Conventional GA		Branch-and-price	
Variation		mean	SD	mean	SD	mean	SD	mean	SD
Task deletion – Number of tasks	1	0.0009	0.0037	9.5308	0.0423	9.5312	0.0373	0.0025	0.0054
	2	0.0013	0.004	11.0164	0.0537	11.0232	0.0518	0.0065	0.0065
	3	0.0013	0.0042	12.4531	0.05	12.4398	0.0491	0.0147	0.0066
	4	0.0016	0.0047	13.8258	0.066	13.8103	0.0487	0.0341	0.008
	5	0.0016	0.0047	15.3255	0.1074	15.3068	0.0764	0.0954	0.0217
	6	0.0016	0.0046	16.6686	0.096	16.6834	0.0757	0.3292	0.0761
	7	0.0019	0.0048	18.0936	0.0956	18.088	0.081	1.4009	0.3368
	8	0.0022	0.005	19.4686	0.09	19.4727	0.0959	8.0993	1.9977
	9	0.0025	0.0057	20.8409	0.14	20.8148	0.1726	55.4437	18.5293
Task deletion – Number of agents	1	0.0019	0.0025	16.1969	0.0761	16.1922	0.0668	4.4595	1.3539
	2	0.0022	0.0052	18.0362	0.1354	18.0242	0.1025	7.8452	2.4496
	3	0.0022	0.0054	19.7478	0.0845	19.7827	0.1194	10.9295	3.3378
	4	0.0025	0.0056	21.3566	0.1015	21.3315	0.0883	14.6456	4.6009
	5	0.0025	0.0056	23.0555	0.1043	23.0463	0.1046	14.9673	4.0564
	6	0.0028	0.0031	24.9809	0.1288	24.9546	0.1051	17.9851	5.5919
	7	0.0029	0.0032	26.8527	0.108	26.8028	0.166	20.6745	6.8098
	8	0.003	0.0028	28.6156	0.13	28.6224	0.1683	21.7446	7.8269
	9	0.0033	0.0036	30.3814	0.1533	30.3969	0.1446	24.179	8.1451
	10	0.0031	0.006	32.1636	0.1524	32.1096	0.1326	25.949	8.2305

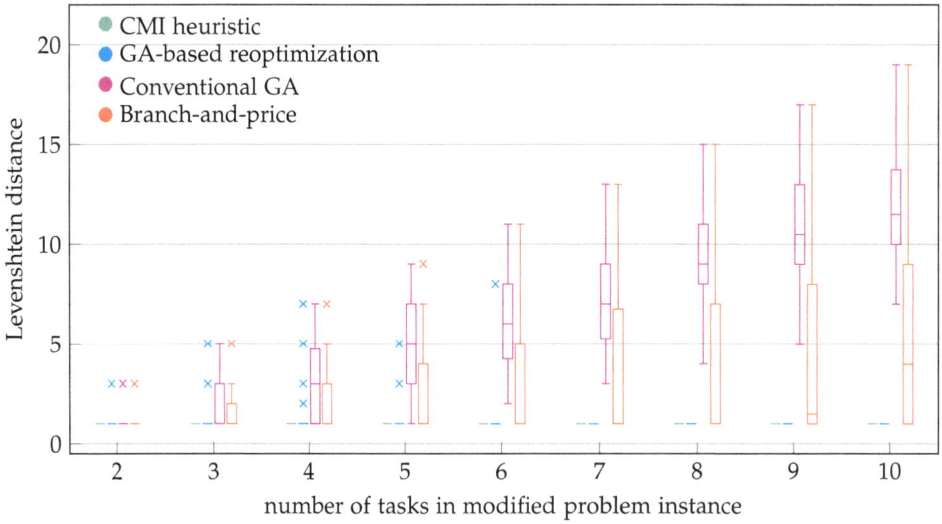

Figure 5.6: Evaluation scenario 1: Levenshtein distances for the modification of task insertion over problem instances with different numbers of tasks. Detailed values are given in Table C.3 in Appendix C.1.

BnP approach grow approximately linearly with the numbers of tasks and vary comparatively more than the solutions of the reoptimization approaches. The Levenshtein distances of the conventional GA grow the most with the numbers of tasks, with the Levenshtein distances being spread between 7 and 19, having a median of 11.5 and an average of 11.82 (see average values given in Table C.3 in Appendix C.1) for problem instances with ten tasks. The Levenshtein distances of the solutions generated by the BnP approach for the modified problem instances with ten tasks have a median of 4.0, an average value of 6.34 and are spread from 1 to 19 over all 50 instances.

The results on the Levenshtein distances of the solutions generated by the different solution approaches for the modification of task insertion over the varying number of agents within the problem specifications is given in Figure 5.7. For all numbers of agents investigated, the Levenshtein distances of the GA-based reoptimization framework equal the ones of the CMI, i. e. 1, for all problem instances. The Levenshtein distances of the solutions generated by the conventional GA are relatively stable for the problem instances with two to ten agents. They vary between 3 and 15, having an average value between 8.64 and 10.76 (see average values given in Table C.3 in Appendix C.1). For the exact solutions generated by the BnP approach, the Levenshtein distances reach values between 1 and 15 for problem instances with up to five agents. For problem instances with more agents, the Levenshtein distances decrease again to being spread from 1 to 5 with an average value of 3.08 for the problem instances with ten agents.

In Table 5.9, the mean values and corresponding standard deviations of the Levenshtein distances of the solutions generated by the different solution approaches for the modification of task deletion are given. By definition, the routings of the solutions generated by the TDH differ from the initial routings by the deletion of one task (see Algorithm 2). Consequently, the Levenshtein distance of the TDH solutions to the initial solution are equal to 1, independently of the number of tasks or agents considered within the modified problem instance. The solutions generated by the GA-based reoptimization framework have a Levenshtein distance of 1 for all problem specifications with varying numbers of agents and for all problem instances of the problem specifications with seven or more tasks. For the problem instances with fewer tasks, the average Levenshtein distance is slightly higher, ranging between 1.12 and 1.90. Since the conventional GA and the BnP approach solve each problem instance independently, i. e. they are optimization and no reoptimization approaches, they show a qualitatively and quantitatively similar dependency on the number of tasks and agents within the modified problem instance as presented above for the modification of task insertion.

5.2.2 Discussion

Overall, both reoptimization heuristics yield very good approximation ratios clearly below $\alpha < 1.02$ which are relatively constant, independently of the number of agents and tasks contained within the modified problem instances. Only for some problem

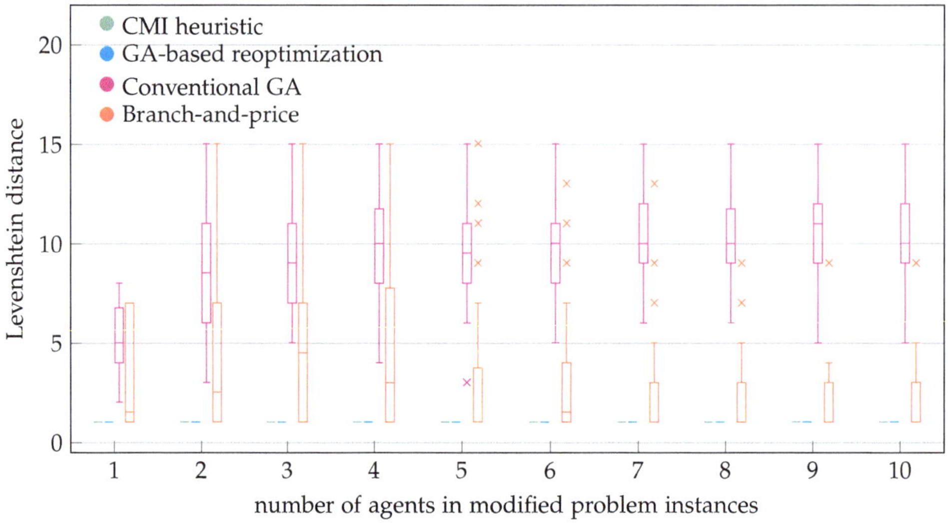

Figure 5.7: Evaluation scenario 1: Levenshtein distances for the modification of task insertion over problem instances with different numbers of agents. Detailed values are given in Table C.3 in Appendix C.1.

instances with up to six tasks, an improvement w. r. t. the approximation ratio of the CMI or TDH solutions is obtained by the additional application of the GA-based reoptimization framework. By definition of the CMI and TDH reoptimization heuristics, their solutions have a fixed Levenshtein distance of 1. Since for some problem instances with up to six tasks, the solutions of the GA-based reoptimization framework deviate from the solutions of the reoptimization heuristics, also the average Levenshtein distances of the corresponding problem specifications deviate from 1 to at most 1.9. Thus, for these problem instances, the average approximation ratios of the CMI and TDH improve by the additional application of the GA-based reoptimization framework, which at the same time increases their Levenshtein distance.

Especially for instances with seven and more tasks, the CMI and the TDH outperform the GA-based reoptimization framework since they generate solutions of the same quality w. r. t. to the approximation ratio in much less computation time and with a constant and low Levenshtein distance of 1. Moreover, taking into consideration the substantially higher computation time required by the GA-based reoptimization framework compared to the reoptimization heuristics, it is questionable whether the potential improvement in approximation ratio for small problem instances with three agents and up to six tasks is worth the loss in responsiveness of the solution approach. For small instances up to this size, the exact computation of a globally optimal solution via the BnP approach requires on average at least about 50 times less computation time and thus might be a better alternative to the GA-based reoptimization framework when putting the focus on solution quality. However, the reduction in computation time by the application of the BnP approach compared to the GA-based

Table 5.9: Evaluation scenario 1: Results on the Levenshtein distance (mean and standard deviation (SD)) the for the modification of task deletion.

Variation			TDH heuristic		GA-based reoptimization		Conventional GA		Branch-and-price	
			mean	SD	mean	SD	mean	SD	mean	SD
Task deletion	Number of tasks	1	1.0	0.0	1.12	0.475	1.12	0.475	1.12	0.475
		2	1.0	0.0	1.16	0.5044	1.58	0.8022	1.3	0.6403
		3	1.0	0.0	1.9	1.7464	2.44	1.6752	2.06	1.7252
		4	1.0	0.0	1.68	1.5677	3.98	2.1771	2.3	1.9723
		5	1.0	0.0	1.44	1.6144	6.1	2.3937	2.72	2.8847
		6	1.0	0.0	1.24	1.68	7.46	2.9679	3.44	3.5505
		7	1.0	0.0	1.0	0.0	9.48	2.9205	3.2	3.7736
		8	1.0	0.0	1.0	0.0	11.52	3.1064	4.4	4.4362
		9	1.0	0.0	1.0	0.0	13.24	2.8111	4.8	5.4699
	Number of agents	1	1.0	0.0	1.0	0.0	6.66	1.38	2.74	2.8622
		2	1.0	0.0	1.0	0.0	9.6	2.569	3.3	3.7216
		3	1.0	0.0	1.0	0.0	10.5	2.816	4.52	4.4955
		4	1.0	0.0	1.0	0.0	10.9	2.6401	4.6	4.5431
		5	1.0	0.0	1.0	0.0	11.4	2.8496	4.5	4.5706
		6	1.0	0.0	1.0	0.0	11.46	2.5156	4.38	4.5028
		7	1.0	0.0	1.0	0.0	12.24	2.4622	4.28	4.4228
		8	1.0	0.0	1.0	0.0	13.04	2.163	3.48	4.3965
		9	1.0	0.0	1.0	0.0	12.32	2.2578	3.32	4.1638
		10	1.0	0.0	1.0	0.0	12.7	2.2561	3.08	3.3873

reoptimization framework for problem instances of this size comes at the cost of potentially higher Levenshtein distances, i. e. less solution stability. Due to the exponential increase in computation time with the numbers of tasks contained in a problem instance, for problems containing nine or more tasks however, the application of the BnP solution approach becomes computationally very expensive. This significant reduction in responsiveness together with the potentially low solution stability makes the BnP approach unsuitable for application in interactive MRTA optimization systems, especially if potentially large problem instances are under consideration.

With the number of agents and tasks contained within a modified problem instance, the calculation time required for the determination of the timing information corresponding to a routing according to (3.5) increases. This is the cause of the approximately linear increase in the computation times required the CMI, TDH and by both GA approaches. The calculation times of the CMI and TDH for all problem specifications are by two orders of magnitude below 300 ms such that these reoptimization heuristics allow for instantaneous interactions [DS15]. This not only holds for the considered problem sizes but also demonstrates their potential for instantaneous interactions for considerably larger problem instances. The higher increase in computation time for the GA approaches is caused by the higher number in different routings evaluated within the course of the GA approaches compared to the reoptimization heuristics. Since the computation times of the GA are mainly determined by the population size and fixed number of 100 iterations performed, there is no significant difference between the conventional GA and the GA-based reoptimization framework. Both GA-based approaches require on average calculation times above 10 s, which makes it improbable for users to stay focused on the interaction if the GA-based approaches were applied in an interactive MRTA optimization system. While this choice of the terminal condition ensures the comparability of the GA approaches w. r. t. solution quality and stability, further optimization of the terminal condition of the GA-based reoptimization framework w. r. t. computation time is possible.

Moreover, the results demonstrate the importance and the influence of the initialization of the GA. The GA-based reoptimization framework clearly outperforms the conventional, randomly initialized GA. This holds both w. r. t. solution quality, i. e. approximation ratio, and solution stability, i. e. Levenshtein distance, even though these GA approaches only differ in their initialization.

In summary, the results of the heuristic reoptimization approaches CMI and TDH scale the best with problem size w. r. t. the combination of solution quality, responsiveness and solution stability and are thus promising for application in interactive MRTA optimization systems. Furthermore, the results of this evaluation scenario indicate that a problem size of three agents and eight tasks is a good compromise between complexity and temporal manageability for the following evaluation scenarios.

After the investigation of the influence of the problem instances' size in this evaluation scenario, the influence of different causes of heterogeneity within the problem instance are evaluated in the following section.

5.3 Evaluation of Different Heterogeneity Levels

The results of the second evaluation scenario and their analysis are presented in this section. As introduced in Section 5.1.3, the modified problem instances contain three agents and eight tasks and no precedence or synchronization constraints. Different levels of agents' capabilities, velocities and tasks' basic durations are considered. The results are presented in the following Section 5.3.1, a discussion of the results is given in Section 5.3.2.

5.3.1 Evaluation Results

The solutions generated by the reoptimization heuristics, the GA-based reoptimization framework, the conventional, randomly initialized GA and the BnP approach w. r. t. their approximation ratios, required calculation times and Levenshtein distances are presented in the following. Different levels of heterogeneity w. r. t. the agents' capabilities, their velocities and the tasks' basic durations (see Section 5.1.3) are analyzed . Since all problem instances investigated are temporally unconstrained, all modifications except for the insertion and deletion of a precedence or synchronization constraint are considered.

Approximation Ratio

The approximation ratios of the solutions generated by the reoptimization heuristics, the GA-based reoptimization framework and the conventional, randomly initialized GA for the problem specifications differing in the levels of the agents' capabilities (c), agents' velocities (v) and tasks' basic durations (d), are given in the following. In Table 5.10, the average values and standard deviations obtained for the modification of task insertion are depicted. The average values of the approximation ratio generated by the CMI heuristic are within a small range of 1.0 to 1.0111 for all heterogeneity levels considered. Out of the different capability levels, level 2 yields by a narrow margin the highest average approximation ratio of 1.0098 with a standard deviation of 0.0306. Compared to the different capability levels, the influence of the different velocity and duration levels on the CMI results is even smaller. Also, heterogeneous velocities or basic task durations in combination with capability level 2 only slightly vary the average approximation ratio.

For the majority of problem specifications, i. e. heterogeneity levels considered, the approximation ratios of the GA-based reoptimization framework are identical to the ones of the CMI. Only for capability level 3 and for the combination of capability level 2 with velocity level 2, the approximation ratios of the GA-based reoptimization framework are a little smaller on average and in standard deviation compared to the CMI. The conventional GA yields exact solutions for the same specifications as the CMI, i. e. homogeneous problems, problems of velocity level 1 and problems of task duration level 1. For the other specifications, the average approximation ratio of the conventional GA ranges up to 1.754 without a clear trend w. r. t. the different heterogeneity levels.

For the modification of task deletion, there is no clear trend of the influence of the different heterogeneity levels on the approximation ratio of the TDH reoptimization heuristic. The average approximation ratios range between about 1.003 and 1.008 for all heterogeneity specifications. For the majority of problem specifications, the approximation ratios of the GA-based reoptimization framework are identical to the approximation ratios of the TDH reoptimization heuristic. The conventional GA yields approximation ratios within a similar range as the ones for the task insertion modification. The detailed results are given in Table C.4 in Appendix C.2.

The approximation ratios generated by the DIH and the INI reoptimization heuristics for the modifications of task position variation and task duration variation are depicted in Table 5.11. The table also indicates the number of instances for which the solution generated by the DIH reoptimization heuristic differs from the one generated by the

Table 5.10: Evaluation scenario 2: Results on the approximation ratio α (mean and standard deviation (SD)) for the modification of task insertion.

Mod.	Level			CMI heuristic		GA-based reoptimization		Conventional GA	
	c	v	d	mean	SD	mean	SD	mean	SD
Task insertion	0	0	0	1.0	0.0	1.0	0.0	1.0	0.0
	1	0	0	1.0058	0.0122	1.0058	0.0122	1.3223	0.0884
	2	0	0	1.0098	0.0306	1.0098	0.0306	1.2784	0.1028
	3	0	0	1.0049	0.0214	1.0019	0.0058	1.0882	0.0812
	0	1	0	1.0	0.0	1.0	0.0	1.0	0.0
	0	2	0	1.0012	0.0036	1.0012	0.0036	1.7153	0.2071
	2	1	0	1.0081	0.0221	1.0081	0.0221	1.28	0.1031
	2	2	0	1.0111	0.0346	1.0102	0.0312	1.3551	0.1509
	0	0	1	1.0	0.0	1.0	0.0	1.0	0.0
	0	0	2	1.0051	0.0163	1.0051	0.0163	1.2924	0.0818
	2	0	1	1.0094	0.0279	1.0094	0.0279	1.3155	0.0943
	2	0	2	1.0084	0.0242	1.0084	0.0242	1.2757	0.0884
	0	2	1	1.0012	0.0035	1.0012	0.0035	1.7538	0.164
	0	2	2	1.0009	0.0027	1.0009	0.0027	1.5848	0.17
	2	2	1	1.0097	0.0294	1.0097	0.0294	1.3511	0.1499
	2	2	2	1.0083	0.0263	1.0083	0.0263	1.3686	0.1403

INI heuristic. Furthermore, considering only the problem instances for which the solutions differ, the average improvement of the approximation ratio of the DIH solutions in comparison to the INI solutions is depicted.

For the modification of task position variation, the average approximation ratios generated by the DIH heuristic range between 1.0032 and 1.0105 for the different heterogeneity specifications with standard deviations between 0.0086 and 0.0226. The average approximation ratios of the solutions generated by the INI heuristic range between 1.0229 and 1.0632 with standard deviations between 0.0440 and 0.0751. Clear influences of the different capability, velocity and duration levels cannot be identified for any of the two heuristics. Within all problem specifications, the DIH and INI solutions differ for at least 17 and up to 33 out of the 50 problem instances. The average improvement of the DIH solutions differing from the INI solutions range between 4.93% and 10.44% for the different problem specifications. As given in Table C.5 in Appendix C.2, for the GA-based reoptimization framework initialized with the DIH heuristic, hardly any improvement of the DIH approximation ratios is obtained. For the GA-based reoptimization framework initialized with the INI heuristic, small improvements compared to the INI solution are generated for some heterogeneity levels.

In comparison to this, for the modification of task duration variation, the average approximation ratios as well as their standard deviations of the solutions generated by the INI and the DIH heuristic are identical for 8 out of the 16 of heterogeneity specifications. For only up to 3 out of 50 problem instances within the specifications, the DIH

Table 5.11: Evaluation scenario 2: Results on the approximation ratio α (mean and standard deviation (SD)) for the modifications of task position variation and task duration variation for the DIH and the INI reoptimization heuristics. Additionally, the number of instances for which the solution generated by the DIH reoptimization heuristic differs from the one generated by the INI heuristic is given. Furthermore, the last column (impr.) gives the average improvement of the approximation ratio of the DIH over the INI solutions (only considering the problem instances, for which the solutions differ).

Mod.	Level c	Level v	Level d	DIH heuristic mean α	DIH heuristic SD	INI heuristic mean α	INI heuristic SD	DIH differs from INI instances	DIH differs from INI impr.
Task position variation	0	0	0	1.0067	0.0146	1.0571	0.0697	30 of 50	8.40%
	1	0	0	1.0096	0.0188	1.0632	0.0751	28 of 50	9.56%
	2	0	0	1.0105	0.0226	1.0471	0.0681	19 of 50	9.64%
	3	0	0	1.0058	0.0172	1.0229	0.0526	15 of 50	5.70%
	0	1	0	1.0062	0.0119	1.0513	0.0555	32 of 50	6.31%
	0	2	0	1.005	0.0121	1.0486	0.0554	33 of 50	5.97%
	2	1	0	1.0064	0.0151	1.0473	0.0699	19 of 50	10.44%
	2	2	0	1.004	0.0085	1.0359	0.0585	17 of 50	9.07%
	0	0	1	1.0067	0.0144	1.0571	0.0698	30 of 50	7.33%
	0	0	2	1.0055	0.0117	1.0458	0.0568	31 of 50	5.80%
	2	0	1	1.0102	0.023	1.0475	0.0707	20 of 50	8.00%
	2	0	2	1.0087	0.0221	1.0389	0.0674	20 of 50	6.49%
	0	2	1	1.005	0.0121	1.0486	0.0548	33 of 50	5.93%
	0	2	2	1.0042	0.0101	1.0397	0.044	33 of 50	4.93%
	2	2	1	1.0047	0.0106	1.0371	0.0582	17 of 50	8.36%
	2	2	2	1.0032	0.0086	1.0308	0.0503	18 of 50	6.84%
Task duration variation	0	0	0	1.0	0.0	1.0	0.0	1 of 50	0.00%
	1	0	0	1.0012	0.0059	1.0012	0.0059	0 of 50	0.00%
	2	0	0	1.0007	0.0047	1.0007	0.005	1 of 50	0.21%
	3	0	0	1.0006	0.0039	1.0013	0.0062	5 of 50	0.70%
	0	1	0	1.0	0.0	1.0	0.0	0 of 50	0.00%
	0	2	0	1.0	0.0	1.0	0.0	0 of 50	0.00%
	2	1	0	1.0	0.0	1.0001	0.0007	2 of 50	0.26%
	2	2	0	1.0001	0.0006	1.0003	0.0018	2 of 50	0.60%
	0	0	1	1.0	0.0	1.0	0.0	1 of 50	0.00%
	0	0	2	1.0	0.0	1.0	0.0	1 of 50	0.00%
	2	0	1	1.0005	0.0033	1.0005	0.0036	1 of 50	0.22%
	2	0	2	1.0003	0.0019	1.0003	0.0022	1 of 50	0.21%
	0	2	1	1.0	0.0	1.0	0.0	0 of 50	0.00%
	0	2	2	1.0	0.0	1.0	0.0	0 of 50	0.00%
	2	2	1	1.0003	0.0018	1.0005	0.0025	2 of 50	0.62%
	2	2	2	1.0	0.0	1.0005	0.0026	3 of 50	0.88%

and INI solutions differ. The corresponding improvement in approximation ratio of the differing DIH solutions is on average at most 0.88% compared to the corresponding INI solutions. For both heuristics, the average approximation ratios are close to optimal ranging between 1.0 and 1.0012, and also for this modification no trends w. r. t. the heterogeneity levels can be identified for the approximation ratios. The corresponding GA-based reoptimization frameworks yield the same approximation ratios as the heuristics they are initialized with (see Table C.6 in Appendix C.2).

Also for the modifications of agent capability variation and agent velocity variation, no clear influences of the different capability, velocity and duration levels can be identified. The average approximation ratios of the INI heuristic range between 1.0 and 1.0435 for the agent capability variation and between 1.0290 and 1.1318 for the agent velocity variation. Only slight improvements of these values are obtained for individual specifications by the application of the GA-based reoptimization framework. The detailed results are given in Table C.7 in Appendix C.2.

Calculation Time

The calculation times required by the CMI and the TDH reoptimization heuristics to solve the task insertion and task deletion problem instances of different heterogeneity levels are within the same magnitude as their calculation time required to solve homogeneous problem instances of the same size. The detailed results are given in Table C.8 in Appendix C.2. For the CMI, the average calculation time ranges between 1.8 ms and 3.2 ms, for the TDH it ranges between 1.3 ms and 4.4 ms. There is no clear trend on how which kind of heterogeneity influences these calculation times. Furthermore, also the average calculation times required by the GA-based reoptimization frameworks and the conventional, randomly initialized GA are relatively stable around about 19.5 s to 20.3 s, i. e. they differ from the calculation times required by the reoptimization heuristics approximately by a factor of 10^4.

The only solution approach for which the calculation times for solving the task insertion and task deletion problem instances are clearly effected by the heterogeneity level is the exact BnP approach. The effects are the same as for the modifications of task position variation and task duration variation, for which the results on the average calculation time and its standard deviation for the DIH heuristic, the INI heuristic and the BnP approach are given in Table 5.12.

For the exact BnP approach, the average computation time needed to solve homogeneous problem instances for the modification of task position variation is about 7.96 s. For capability level 1, where the capabilities of the agents to perform the tasks vary, but are all positive (see Table 5.2), the average computation time increases slightly to about 8.65 s. To solve problem instances of capability level 2, in which two of the three agents have no capability to perform one task type, the average calculation time of the BnP approach decreases to about 3.15 s. The smallest average calculation time of about

Table 5.12: Evaluation scenario 2: Results on the calculation time in seconds (mean and standard deviation (SD)) for the modifications of task position variation and task duration variation for the DIH and the INI heuristics.

Mod.	Level			DIH heuristic		INI heuristic		Branch-and-price	
	c	v	d	mean	SD	mean	SD	mean	SD
Task position variation	0	0	0	0.0048	0.0068	0.0034	0.0063	7.9584	2.3851
	1	0	0	0.0036	0.0061	0.0018	0.0047	8.6492	2.6216
	2	0	0	0.0031	0.0062	0.0032	0.0061	3.1495	1.0826
	3	0	0	0.0041	0.0068	0.0024	0.0055	0.7105	0.8992
	0	1	0	0.0059	0.0124	0.0025	0.0057	7.2826	2.0259
	0	2	0	0.0038	0.0052	0.0032	0.0106	5.1822	1.6548
	2	1	0	0.004	0.0065	0.0024	0.0053	3.376	1.1757
	2	2	0	0.0034	0.0058	0.0028	0.0055	3.1101	1.1323
	0	0	1	0.0049	0.007	0.0024	0.0053	8.1114	2.4378
	0	0	2	0.0046	0.0069	0.002	0.005	9.7915	3.3737
	2	0	1	0.006	0.0109	0.002	0.0051	3.4315	1.2444
	2	0	2	0.0047	0.007	0.002	0.0051	4.3473	1.4569
	0	2	1	0.0039	0.0062	0.0032	0.0058	5.4368	1.7139
	0	2	2	0.0043	0.0067	0.0011	0.0036	6.6701	2.2263
	2	2	1	0.0042	0.0067	0.0022	0.0054	3.2069	1.5318
	2	2	2	0.0056	0.0073	0.0016	0.0047	4.1095	1.209
Task duration variation	0	0	0	0.0059	0.0075	0.0014	0.0038	7.5067	2.2915
	1	0	0	0.0042	0.0066	0.0028	0.0056	8.5191	2.3673
	2	0	0	0.0037	0.0065	0.0031	0.006	3.6616	1.3735
	3	0	0	0.0046	0.007	0.0019	0.0046	0.7378	0.9946
	0	1	0	0.0052	0.0071	0.0018	0.0048	7.143	2.2115
	0	2	0	0.0034	0.0044	0.002	0.0039	5.0668	1.4212
	2	1	0	0.0053	0.0073	0.0015	0.0044	3.5769	1.6092
	2	2	0	0.0046	0.0067	0.002	0.0048	3.3709	1.4947
	0	0	1	0.0041	0.0063	0.0016	0.0047	7.5589	2.0451
	0	0	2	0.0041	0.0067	0.0015	0.0045	9.0524	3.0947
	2	0	1	0.0056	0.0072	0.0017	0.0046	3.7682	1.3325
	2	0	2	0.0049	0.0069	0.0018	0.0045	4.5276	1.5566
	0	2	1	0.0051	0.0103	0.0026	0.0054	5.1572	1.4974
	0	2	2	0.0053	0.0071	0.0013	0.0042	6.5357	2.1128
	2	2	1	0.0047	0.0069	0.0024	0.0055	3.2877	1.3111
	2	2	2	0.0058	0.0122	0.0016	0.0047	4.0817	1.6441

0.71 s, i. e. about 10% of the average calculation time for homogeneous problem instances, is needed to solve problem instances of capability level 3. In capability level 3 two out of the three agents are only capable to perform one task type (see Table 5.2). An increase in velocity levels also leads to a slight decrease in the average calculation time required by the BnP approach, which reduces to about 5.18 s for velocity level 2. Contrary to this, an increase in duration level slightly increases the average calculation time, which is equal to about 9.79 s for basic task duration level 2. However, also for the combinations of the different heterogeneity levels, the capability level has the highest impact on the required calculation time. The same dependencies can be observed for the BnP approach applied to the modified problem instances corresponding to the modifications of task duration variation, task insertion and task deletion (see Table C.8 in Appendix C.2).

For the modifications of task position variation and task duration variation both reoptimization heuristics, i. e. INI and DIH, require average calculation times that are relatively stable for all heterogeneity levels. As depicted in Table 5.12, for the INI heuristic, they vary between 1.1 ms and 3.4 ms for both modifications with standard deviations between 3.6 ms and 10.6 ms. The average calculation times required by the DIH are approximately twice as high, varying between 3.1 ms and 6.0 ms for both modifications with similar standard deviations. For the modifications of agent capability variation and agent velocity variation, the average calculation times for the INI heuristic are approximately within the same range (see Table C.11 in Appendix C.2). The calculation times required by the GA-based reoptimization framework and the conventional GA are relatively stable around 20 s, independently of the modification under consideration (see Tables C.8 to C.11).

Levenshtein Distance

For the modifications of task insertion and task deletion, the Levenshtein distances of the solutions generated by the CMI and the TDH, respectively, are, by definition of the reoptimization heuristics, equal to one. The application of the GA-based reoptimization framework slightly increases the average Levenshtein distance to a maximum of 1.18 for two out of 16 heterogeneity specifications for the modification of task insertion and for one out of 16 heterogeneity specifications for the modification of task deletion. For both modifications, the exact solutions generated by the BnP approach have an average Levenshtein distance varying between 2.26 and 5.22 with no clear trend regarding the corresponding heterogeneity specification. For all heterogeneity specifications considered, the results generated by the conventional, randomly initialized GA yield the highest Levenshtein distances for both modifications, which range up to an average value of 11.14. The detailed results of the Levenshtein distances for the modifications of task insertion and task deletion are given in Table C.12 in Appendix C.2.

The results on the Levenshtein distances of the solutions generated for the problem instances corresponding to the modifications of task position variation and task duration variation are depicted for the DIH, the INI and the BnP approach in Table 5.13.

By definition of the INI heuristic (see Algorithm 3), the Levenshtein distances of the corresponding solutions are equal to 0. For the modification of task position variation, the DIH heuristic yields solutions with average Levenshtein distances ranging between 0.6 and 1.32 with a standard deviation of about 0.92 to 0.99. An increase in the agent's capability levels decreases the resulting Levenshtein distances. The average Levenshtein distance of the homogeneous problem specification (i. e. capability, velocity and duration level 0) equals 1.2 which reduces to 0.76 for problems of capability level 2 and to 0.6 for problems of capability level 3. Also, the comparison of different combinations of velocity and duration levels with either capability level 0 or capability level 2 reveals almost up to factor 2 smaller average Levenshtein distances for combinations with capability level 2. No clear influence of different velocity and duration levels can be identified for the solutions generated by the DIH for the modification of task position variation. When solving problems of the modification task duration variation, the average Levenshtein distances generated by the DIH vary between 0.0 and 0.12 with standard deviations in a range of 0.0 to about 0.48. For problems corresponding to this modification, neither an influence of the capability level nor of the velocity or duration level can be identified. The solutions generated by the exact BnP approach have an average Levenshtein distance ranging between 2.32 and 5.62 for the different heterogeneity levels when applied to task position variation problems. In comparison to this, the average Levenshtein distances of the BnP solutions to task duration variation problems are much smaller and range between 0.16 and 1.56 for the different heterogeneity levels. For both modifications, no clear influence of the capability, velocity and duration levels on the Levenshtein distances of the BnP solutions can be identified.

For both modifications of task position variation and task duration variation, the Levenshtein distances of the solutions generated by the GA-based reoptimization frameworks initialized with DIH or INI, respectively, are equal to the Levenshtein distances of the DIH or INI solutions for the majority of heterogeneity levels (see Tables C.13 and C.14 in Appendix C.2). This also holds for the GA-based reoptimization framework initialized with INI when applied to problems corresponding to the modifications of agent capability variation and agent velocity variation. The overview of the average Levenshtein distances and their standard deviations for the modifications of agent capability variation and agent velocity variation are given in Table C.15 in Appendix C.2. For the modification of agent capability variation, the Levenshtein distances of the exact solutions generated by the BnP approach vary between 0.84 and 3.98, the range of the average Levenshtein distances of the BnP solutions to problems of the modification agent velocity variation is given by 1.72 to 7.76. No clear influence of the different heterogeneity levels on the corresponding Levenshtein distances of the BnP solutions can be identified for the agent capability variation problems. For agent velocity variation problems, higher capability levels reduce the corresponding Levenshtein distances for BnP solutions from an average of 7.56 for homogeneous problems to 2.06 for capability level 3. Also in combination with other velocity and duration levels, capability level 2

Table 5.13: Evaluation scenario 2: Results on the Levenshtein distance (mean and standard deviation (SD)) for the modifications of task position variation and task duration variation for the DIH and the INI heuristics.

Mod.	Level			DIH heuristic		INI heuristic		Branch-and-price	
	c	v	d	mean	SD	mean	SD	mean	SD
Task position variation	0	0	0	1.2	0.9798	0.0	0.0	4.3	4.6615
	1	0	0	1.12	0.9928	0.0	0.0	5.62	4.0541
	2	0	0	0.76	0.9708	0.0	0.0	4.22	3.8538
	3	0	0	0.6	0.9165	0.0	0.0	2.32	2.2221
	0	1	0	1.28	0.96	0.0	0.0	4.68	4.5758
	0	2	0	1.32	0.9474	0.0	0.0	3.5	3.189
	2	1	0	0.76	0.9708	0.0	0.0	2.58	3.0925
	2	2	0	0.68	0.9474	0.0	0.0	4.46	3.8585
	0	0	1	1.2	0.9798	0.0	0.0	4.92	4.4937
	0	0	2	1.24	0.9708	0.0	0.0	4.54	4.5879
	2	0	1	0.8	0.9798	0.0	0.0	4.06	3.8803
	2	0	2	0.8	0.9798	0.0	0.0	3.26	3.4975
	0	2	1	1.32	0.9474	0.0	0.0	3.44	3.0735
	0	2	2	1.32	0.9474	0.0	0.0	3.54	3.1062
	2	2	1	0.68	0.9474	0.0	0.0	3.78	3.684
	2	2	2	0.72	0.96	0.0	0.0	3.06	3.3431
Task duration variation	0	0	0	0.04	0.28	0.0	0.0	0.28	1.3862
	1	0	0	0.0	0.0	0.0	0.0	1.26	3.7138
	2	0	0	0.04	0.28	0.0	0.0	0.68	2.4855
	3	0	0	0.2	0.6	0.0	0.0	0.56	1.7339
	0	1	0	0.0	0.0	0.0	0.0	0.16	1.12
	0	2	0	0.0	0.0	0.0	0.0	0.32	1.5677
	2	1	0	0.08	0.3919	0.0	0.0	0.16	0.88
	2	2	0	0.08	0.3919	0.0	0.0	0.78	2.0522
	0	0	1	0.04	0.28	0.0	0.0	0.16	0.88
	0	0	2	0.04	0.28	0.0	0.0	0.24	1.0307
	2	0	1	0.04	0.28	0.0	0.0	1.56	3.2812
	2	0	2	0.04	0.28	0.0	0.0	1.28	3.0136
	0	2	1	0.0	0.0	0.0	0.0	0.76	2.294
	0	2	2	0.0	0.0	0.0	0.0	0.48	1.8999
	2	2	1	0.08	0.3919	0.0	0.0	1.2	2.735
	2	2	2	0.12	0.475	0.0	0.0	0.92	2.5047

yields average Levenshtein distances for the BnP solutions that are up to almost factor 3 lower compared to homogeneous capabilities of level 0.[37]

The Levenshtein distances of the conventional, randomly initialized GA range approximately between 5.4 and 11.1, independently of the problem modification and the heterogeneity level (see Tables C.12 to C.15 in Appendix C.2).

A discussion of the presented results on the approximation ratios, calculation times and Levenshtein distances resulting for problems of different heterogeneity levels is given in the following section.

5.3.2 Discussion

Within the second evaluation scenario, the influence of different causes of heterogeneity on the solutions generated by the different reoptimization heuristics, the corresponding GA-based reoptimization frameworks, the conventional, randomly initialized GA and by the BnP approach are analyzed.

Overall, no distinct influence of the different agents' capability and velocity levels and the different tasks' basic duration levels considered on the approximation ratios resulting from any of the investigated solution approaches for the considered modifications can be identified. Only the capability level has a slight influence on the approximation ratios for the modifications of task insertion. Capability level 2, in which agents differ in their capabilities in performing different task types and two agents are only capable of performing two out of three task types (see Table 5.2), slightly increases the approximation ratios compared to problem instances with homogeneous capabilities. However, the approximation ratios of all reoptimization approaches are as close to the optimum as desired ($\alpha < 1.2$), and clearly outperform the optimization approach using the conventional, randomly initialized GA.

Neither can a general trend of the heterogeneity levels influencing the resulting Levenshtein distances of the solution approaches be identified. Only for the modification of task position variation, higher capability levels with more unambiguous agent-task allocations, slightly decrease the average Levenshtein distances of the DIH reoptimization heuristic. This is probably caused by the reduced number of alternative routes in which the task with the altered position can be inserted due to the increased chance of incapable alternative agents.

Furthermore, also the calculation times required by the CMI, TDH, DIH, and INI reoptimization heuristics, as well as by the GA-based approaches are not clearly affected by

[37] An average Levenshtein distance of 5.10 is given for problems of capability level 0, velocity level 2 and duration level 2 compared to an average Levenshtein distance of 1.72 for problems of capability level 2, velocity level 2 and duration level 2.

the different heterogeneity levels. The additional application of the GA-based reoptimization framework compared to only applying the respective reoptimization heuristic, only slightly improves the resulting approximation ratios of some problem instances. Thus, it is questionable whether spending the by orders of magnitude higher additional calculation time is worth the comparatively small potential improvement in solution quality. Only the calculation time required by the exact BnP approach is affected by the heterogeneity levels. It reduces considerably with an increase in unambiguous agent-task allocations which result from only some or even only one agent having capabilities greater zero to perform certain tasks. For capability level 3, in which two of the three agents are only capable to perform one of the three task types, the calculation time reduces to about 10% of the calculation time required for the homogeneous problem instances. However, the calculation times required by the BnP approach are still by several orders of magnitude higher compared to the respective reoptimization heuristics. Moreover, if the capabilities of the agents to perform the tasks differ, but are all greater than zero, i. e. no agent-task-allocation can be excluded, the calculation time required by the BnP increases slightly compared to the problem specifications with homogeneous capabilities.

The comparison of the different problem modifications is especially interesting for the modifications of task position variation and task duration variation. For both modifications, both the DIH and the INI reoptimization heuristic are applicable. While the INI heuristic keeps the initial routing, the DIH tries to improve the initial routing by deleting and reinserting the task of which the position or duration has been modified (see Algorithm 5). This additional effort results in calculation times that are approximately twice as high for the DIH compared to the INI heuristic. However, the DIH still generates results within milliseconds and can thus be regarded as having a good responsiveness when applied in interactive MRTA optimization systems [DS15]. The application of the DIH results in much higher improvements of the resulting approximation ratios compared to the INI heuristic for the problem instances of the modification task position variation compared to the modification of task duration variation. On average over all problem instances of all heterogeneity levels considered, the DIH yields an approximation ratio that improves the approximation ratio of the INI heuristic by 3.6847%. For the task duration variation problem instances, this average improvement is only 0.0106% and thus almost 350 times smaller than for the task position variation problem instances. Obviously, the improvement in solution quality by the DIH comes at the cost of slightly higher Levenshtein distances, i. e. reduced solution stability. However, by the definition of the DIH, in which at most one task is deleted in one position of a route and inserted in another position of any route, the resulting Levenshtein distance can only equal zero or two, and is thus always within the acceptable range. Hence, no arbitrarily low solution stability can result by the application of the DIH compared to the INI heuristic that by definition always yields the optimum Levenshtein distance of zero, since the initial routing is not altered. Consequently, the slight decrease in responsiveness of the DIH compared to the INI heuristic, yields much higher improvements in solution quality for the problem instances considered for the modification of task position variation compared to the ones of the modification task duration variation. These improvements slightly reduce the solution stability, which is however bounded above by a Levenshtein distance of two.

The comparison of the DIH and the INI heuristic for task position and task duration variation problems implies that the variation of a task position as conducted within this evaluation has a higher influence on the optimal routing compared to the conducted basic task duration variation. This assumption is supported by the Levenshtein distances of the exact solutions generated by the BnP approach, which are on average 5.77 times higher for the problem instances corresponding to the modification of task position variation compared to those corresponding to the modification of task duration variation. Thus, the task position variation has a higher impact on the optimal solution than the task duration variation. Based on this measure, the influence of the different modifications considered within this evaluation scenario on the variation of the optimal modified solution compared to the initial solution is as follows: The modification of duration variation varies the optimal solution the least with an overall average Levenshtein distance of 0.675 of the BnP solutions, followed by the modification of agent capability variation that yields an average Levenshtein distance of 2.035 of all BnP solutions. The modifications of task insertion, task deletion and task position variation show similar influences with average BnP Levenshtein distances of 3.424, 3.769 and 3.893. The highest average Levenshtein distance of all BnP solution of 4.763 results for the modification of agent velocity variation, which thus influences the optimal solution the most.

In summary, the proposed reoptimization heuristics outperform the other evaluated solution approaches w. r. t. the combination of solution quality, responsiveness and solution stability independent of the considered heterogeneity level. Furthermore, in this evaluation scenario, the DIH is superior to the INI reoptimization heuristic, especially for the modification of task position variation.

However, so far only temporally unconstrained problem instances have been analyzed. Thus, in the following section, the influence of the presence of precedence and synchronization constraints within the problem instances are evaluated. In addition to the modifications considered within the heterogeneity investigation, also the modifications of precedence and synchronization constraint insertion as well as precedence and synchronization constraint deletion are assessed.

5.4 Evaluation of Problem Instances with Temporal Constraints

In this section, the results of the third evaluation scenario are presented. As introduced in Section 5.1.4, the modified problem instances contain three agents and eight tasks. They are homogeneous w. r. t. agent capabilities and basic task durations and vary in the number of precedence and synchronization constraints considered within the initial problem instances (see Section 5.1.4). All modifications analyzed in Chapter 3 are considered within this evaluation scenario (i. e. task insertion, task deletion, task position variation, task duration variation, agent capability variation, agent velocity variation, precedence constraint insertion, synchronization constraint insertion, precedence constraint deletion and synchronization constraint deletion). The results on the respective approximation ratios, calculation times and Levenshtein distances are presented in the upcoming section, followed by a discussion thereof in Section 5.4.2.

5.4.1 Evaluation Results

The results w. r. t. the approximation ratios of the solutions generated by the different reoptimization and optimization approaches of the evaluation of the influence of precedence and synchronization constraints contained within the initial problem instance are presented in the following. Subsequently, the required calculation times and the corresponding Levenshtein distances are given.

Approximation Ratio

The mean values and corresponding standard deviations of the approximation ratios generated for the problem instances corresponding to the modifications of task insertion and task deletion are given in Table 5.14. For the modification of task insertion, the average approximation ratios of the solutions generated by the CMI heuristic vary between 1.0 and 1.0155, for the modification of task deletion and the corresponding TDH, they range from 1.0083 to 1.0351. For both modifications and corresponding reoptimization heuristics, the approximation ratios increase most with the number of synchronization constraints contained within the problem instance. The approximation ratios of the problem instances with one synchronization constraint are already higher than the ones of the problem instances considering two precedence constraints. These values increase further for both modifications when two synchronization constraints are considered. For both modifications, the additional application of the GA-based reoptimization framework yields either none or only small improvements of at most 0.04% of the resulting average approximation ratios. The approximation ratios generated by the conventional, randomly initialized GA are by orders of magnitude higher and range between 1.23 and 1.36 for both modifications for problem instances containing temporal constraints.

In contrast to this, no clear influence of the number of precedence and synchronization constraints contained within the problem instances can be observed for the application of the INI heuristic to modified problem instances corresponding to the modifications of agent capability variation and agent velocity variation. The corresponding results are depicted in Table C.16 in Appendix C.3. While the additional application of the GA-based reoptimization framework on average does not yield improved approximation ratios for the majority of problem specifications for the modification of agent capability

Table 5.14: Evaluation scenario 3: Results on the approximation ratio α (mean and standard deviation (SD)) for the modifications of task insertion and task deletion.

Mod.	Constraints in $\mathcal{I}$ sync.	prec.	Reoptimization heuristic mean	SD	GA-based reoptimization mean	SD	Conventional GA mean	SD
Task insertion	0	0	1.0	0.0	1.0	0.0	1.0	0.0
	0	1	1.0043	0.0187	1.0043	0.0187	1.2763	0.1165
	0	2	1.0043	0.0104	1.0043	0.0104	1.3378	0.1289
	1	0	1.0126	0.0208	1.0125	0.0206	1.229	0.1009
	2	0	1.0155	0.0213	1.0155	0.0213	1.2394	0.1121
	1	1	1.0153	0.0285	1.0153	0.0285	1.3007	0.132
Task deletion	0	0	1.0083	0.0147	1.0083	0.0147	1.3576	0.1097
	0	1	1.0076	0.0133	1.0076	0.0133	1.345	0.1045
	0	2	1.0107	0.0228	1.0107	0.0228	1.3422	0.1412
	1	0	1.0292	0.0444	1.0292	0.0444	1.2554	0.1124
	2	0	1.0351	0.06	1.0347	0.0578	1.2543	0.1102
	1	1	1.0295	0.0417	1.0295	0.0417	1.2924	0.129

variation, an average improvement ranging from 0.3% to 1.8% for the different problem specifications results for the modification of agent velocity variation.

For the modifications of task position variation and task duration variation, the results on the approximation ratios generated by the application of the DIH and the INI reoptimization heuristic are depicted in Table 5.15. Additionally, the number of problem instances is given, for which the solutions generated by the DIH and INI heuristics differ. For these instances, also the average improvement of the DIH approximation ratio over the one of the INI heuristic is indicated.

As observed within evaluation scenario 2 considering different heterogeneity levels (see Section 5.3.1), the approximation ratios of the solutions generated by INI for problem instances corresponding to the modification of task duration variation are smaller compared to the approximation ratios of the INI solutions for problem instances corresponding to the modification of task position variation. For the problem instances of task duration variation they lie within a range of mean values between 1.0 to 1.0008. For the problem instances of task position variation the average approximation ratios range from 1.0536 to 1.0606. While for the INI results no clear influence of the number and type of temporal constraints contained in the problem instances on the approximation ratios can be observed, the approximation ratios generated by the DIH reoptimization heuristic slightly increase for both modifications with the number of synchronization constraints considered. The average approximation ratio for temporally unconstrained problem instances equals 1.0067 for the modification of task position variation and 1.0 for the modification of task duration variation. These values increase up to 1.0639 and 1.0260 for problem instances with two synchronization constraints.

Table 5.15: Evaluation scenario 3: Results on the approximation ratio α (mean and standard deviation (SD)) for the modifications of task position variation and task duration variation for the DIH and the INI reoptimization heuristic. Additionally, the number of instances for which the solution generated by the DIH reoptimization heuristic differs from the one generated by the INI heuristic is given. Furthermore, the last column (impr.) gives the average improvement of the approximation ratio of the DIH over the INI solutions (only considering the problem instances, for which the solutions differ).

Mod.	Constraints in $\mathcal{I}$		DIH heuristic		INI heuristic		DIH differs from INI	
	sync.	prec.	mean α	SD	mean α	SD	instances	impr.
Task position variation	0	0	1.0067	0.0146	1.0571	0.0697	30 of 50	7.34%
	0	1	1.0302	0.1076	1.0566	0.0696	28 of 50	3.93%
	0	2	1.0227	0.0765	1.0536	0.0727	25 of 50	5.23%
	1	0	1.0421	0.0729	1.0577	0.073	31 of 50	2.16%
	2	0	1.0639	0.1043	1.0606	0.0604	33 of 50	−0.55%
	1	1	1.0393	0.0492	1.0591	0.0821	31 of 50	2.24%
Task duration variation	0	0	1.0	0.0	1.0	0.0	1 of 50	0.00%
	0	1	1.0	0.0	1.0	0.0	1 of 50	0.00%
	0	2	1.0	0.0	1.0	0.0	2 of 50	0.00%
	1	0	1.0151	0.0303	1.0001	0.0004	14 of 50	−5.38%
	2	0	1.026	0.0426	1.0008	0.0029	20 of 50	−6.31%
	1	1	1.0187	0.0477	1.0005	0.0022	11 of 50	−8.25%

For the modification of task duration variation, the DIH and INI solutions differ for at most 20 out of the 50 problem instances. For these instances containing synchronization constraints, the approximation ratio of the DIH solutions is on average worse than the approximation ratio of the INI solutions. They have an on average 5.38% to 8.25% higher approximation ratio than the corresponding INI solutions. For the problem instances containing no synchronization constraint however, all solutions of the DIH and INI heuristic have exactly the same approximation ratios and the solutions only differ for one or two out of the 50 problem instances.

For the modification of task position variation, within all problem specifications, the DIH and INI solutions differ for at least half of the 50 problem instances. The DIH improves the approximation ratios of these instances on average by 7.34% within the temporally unconstrained problem specification. Within the problem specifications with one and two precedence constraints, the average improvements amount to 3.93% and 5.23%, respectively. For the problem specifications containing synchronization constraints, the improvement is smaller ranging from a degradation of −0.55% to an improvement of 2.24%.

For both modifications, the conventional GA yields approximation ratios within a similar range as for the modifications of task insertion and task deletion. With both GA-based reoptimization frameworks, only small improvements for some problem specifications can be achieved. These results are depicted in Table C.17 in Appendix C.3.

Within this evaluation scenario, also the modifications of precedence and synchronization constraint insertion and deletion are investigated. The results on the approximation ratios generated by the eDIH for the insertion of a constraint and by the INI reoptimization heuristic for the deletion of a constraint as well as the results of the corresponding GA-based reoptimization frameworks are given in Table 5.16.

For the modification of precedence constraint insertion, the average approximation ratios resulting from the eDIH reoptimization heuristic range between 1.0313 and 1.0425 and no clear influence of the number or type of constraints considered within the initial problem instances can be observed. Also, the additional application of the GA-based reoptimization framework does yield at most small improvements of 0.2% of the average approximation ratios. The average approximation ratios resulting from the eDIH reoptimization heuristic for the problem instances corresponding to the modification of synchronization constraint insertion range from 1.1 to 1.1837 and are thus higher than the ones generated for the modification of precedence constraint insertion for all problem specifications. However, for all problem specifications, an improvement of the average approximation ratios results from the additional application of the GA-based reoptimization framework. The improvement ranges from 0.7% for the initial problem instances with one precedence and one synchronization constraint to 4.7% for the initial problem instances without temporal constraints.

While the INI reoptimization heuristic yields small average approximation ratios between 1.0042 and 1.0196 for the modification of precedence constraint deletion which are not improved by the application of the GA-based reoptimization framework, higher average approximation ratios between 1.1033 and 1.2286 result for the modification of synchronization constraint deletion. The resulting approximation ratios are the highest for initial problem instances with two synchronization constraints. For these problem instances, an average improvement of the approximation ratios of 1.9% results from the additional application of the GA-based reoptimization framework.

Calculation Time

The calculation times required by the different solution approaches to solve the problem instances corresponding to the modifications of task insertion and task deletion are given in Table 5.17. Independently of the numbers and types of constraints considered within the problem instances, the average calculation times required by the CMI and TDH reoptimization heuristics are all below 4 ms and within a similar range as for the problem instances of different heterogeneity levels investigated in the previous evaluation scenario (see Section 5.3.1). However, the average calculation time required by both GA approaches to solve problem instances with precedence or synchronization constraints reduces to approximately half of the time required for problem instances without any temporal constraints. It is the lowest for problem instances with two synchronization constraints, for which on average about 8.8 s to 9.4 s are required. The opposite effect occurs for the problem instances being solved using the

Table 5.16: Evaluation scenario 3: Results on the approximation ratio α (mean and standard deviation (SD)) for the modifications of precedence constraint insertion, synchronization constraint insertion, precedence constraint deletion and synchronization constraint deletion.

Mod.	Constraints in $\mathcal{I}$		Reoptimization heuristic		GA-based reoptimization	
	sync.	prec.	mean	SD	mean	SD
Precedence constr. insertion	0	0	1.0412	0.0459	1.0412	0.0459
	0	1	1.0313	0.0781	1.0302	0.0737
	0	2	1.0316	0.0495	1.0316	0.0495
	1	0	1.0358	0.0732	1.0354	0.0717
	2	0	1.0425	0.0913	1.0404	0.0867
	1	1	1.0252	0.0464	1.0252	0.0464
Synchronization constr. insertion	0	0	1.1837	0.1348	1.1278	0.0786
	0	1	1.1611	0.1434	1.1263	0.0822
	0	2	1.1792	0.1927	1.1443	0.1329
	1	0	1.1	0.0848	1.0863	0.0671
	2	0	1.1162	0.1085	1.1039	0.093
	1	1	1.1021	0.1085	1.0944	0.0842
Precedence constr. deletion	0	0				
	0	1	1.0042	0.0199	1.0042	0.0199
	0	2	1.0179	0.0319	1.0179	0.0319
	1	0				
	2	0				
	1	1	1.0196	0.0352	1.0196	0.0352
Synchronization constr. deletion	0	0				
	0	1				
	0	2				
	1	0	1.1033	0.0799	1.1033	0.0799
	2	0	1.2286	0.1673	1.2052	0.138
	1	1	1.1327	0.0921	1.1289	0.091

exact BnP approach. Every additional temporal constraint increased the average calculation time needed. However, there is a significant difference between precedence and synchronization constraints contained within the problem instances. For example, two precedence constraints increase the average calculation time for the task insertion problem instances to on average 43.2 s from about 7.5 s for the problem instances without temporal constraints. Contrary to this, on average 2159.5 s, i. e. almost 36 min, are required by the BnP approach for problem instances considering two synchronization constraints. The clear increase in standard deviation is within the similar order of magnitude.

The same effects considering the influence of precedence and synchronization constraints on the calculation times required by the GA-based reoptimization framework, the conventional, randomly initialized GA, and the BnP approach can be observed for

Table 5.17: Evaluation scenario 3: Results on the calculation time in seconds (mean and standard deviation (SD)) for the modifications of task insertion and task deletion.

Mod.	Constraints in $\mathcal{I}$ sync.	prec.	Reoptimization heuristic mean	SD	GA-based reoptimization mean	SD	Conventional GA mean	SD	Branch-and-price mean	SD
Task insertion	0	0	0.0018	0.0048	19.551	0.215	19.547	0.121	7.472	2.748
	0	1	0.0031	0.006	10.649	0.283	10.575	0.249	19.469	27.56
	0	2	0.0024	0.0005	9.833	0.306	9.789	0.258	43.157	52.879
	1	0	0.001	0.0037	10.173	0.249	10.113	0.19	410.721	655.749
	2	0	0.0013	0.0042	8.918	0.499	8.835	0.386	2159.463	2974.18
	1	1	0.0025	0.0037	9.433	0.218	9.42	0.212	473.111	366.966
Task deletion	0	0	0.0014	0.0038	19.519	0.183	19.53	0.122	7.745	3.294
	0	1	0.0016	0.0047	12.639	3.651	12.611	3.674	15.473	20.826
	0	2	0.0024	0.0001	10.8	2.29	10.721	2.321	28.77	41.93
	1	0	0.0019	0.0051	12.025	3.813	11.981	3.841	392.979	529.784
	2	0	0.0018	0.0037	9.388	0.865	9.369	0.881	1233.356	1339.11
	1	1	0.0019	0.0051	9.722	0.508	9.668	0.459	403.488	554.367

the modifications of agent capability variation, agent velocity variation, task position variation, task duration variation, precedence constraint deletion and synchronization constraint deletion. The corresponding results are given in Tables C.19 to C.21 in Appendix C.3. These also contain the average calculation times required by the respective reoptimization heuristics which are within a similar range as in the previous evaluation scenario (see Section 5.3.1) and for which no clear influence of the type and number of temporal constraints considered within the problem instances can be observed.

The results on the calculation times required by the different solution approaches for the modifications of precedence and synchronization constraint insertion are given in Table 5.18. The average calculation times required by the eDIH reoptimization heuristic to solve these problem instances range from 12.4 ms to 14.4 ms for the modification of precedence constraint insertion and from 12.9 ms to 17.6 ms for the modification of synchronization constraint insertion. No clear trend w. r. t. the number and type of temporal constraints can be identified. The calculation times required by the GA-based solution approaches and the BnP approach show the same effects as presented above. To solve the problem instances containing three synchronization constraints (synchronization constraint insertion with the initial problem instances containing two synchronization constraints), the average calculation time required by the BnP approach increases to over 106 min.

Levenshtein Distance

The Levenshtein distances of the solutions generated by the different solution approaches to the problem instances of the modifications task insertion and task deletion

Table 5.18: Evaluation scenario 3: Results on the calculation time in seconds (mean and standard deviation (SD)) for the modifications of precedence constraint insertion, synchronization constraint insertion, precedence constraint deletion and synchronization constraint deletion.

Mod.	Constraints in $\mathcal{I}$		eDIH heuristic		GA-based reoptimization		Conventional GA		Branch-and-price	
	sync.	prec.	mean	SD	mean	SD	mean	SD	mean	SD
Precedence constr. insertion	0	0	0.0125	0.0037	10.809	0.246	10.788	0.298	17.8	21.56
	0	1	0.0137	0.006	9.89	0.201	9.839	0.244	43.76	53.43
	0	2	0.0144	0.0042	9.838	0.259	9.803	0.261	76.88	80.64
	1	0	0.0134	0.0054	9.737	0.234	9.73	0.25	539.66	550.01
	2	0	0.0124	0.0062	9.213	0.174	9.205	0.167	1807.73	1872.11
	1	1	0.0147	0.0048	9.699	0.203	9.678	0.234	733.85	630.01
Synchronization constr. insertion	0	0	0.0169	0.0023	10.076	0.244	9.999	0.21	412.75	375.96
	0	1	0.0176	0.0056	9.523	0.201	9.553	0.212	630.2	954.67
	0	2	0.0167	0.0045	9.684	0.202	9.614	0.204	670.29	616.11
	1	0	0.0142	0.0057	9.027	0.128	8.958	0.138	2025.56	1881.62
	2	0	0.0129	0.0075	8.644	0.116	8.639	0.116	6370.99	5513.63
	1	1	0.015	0.0071	9.203	0.224	9.121	0.196	2697.25	2749.53

are similar to the results on the Levenshtein distances within the second evaluation scenario for these modifications (see Section 5.3.1). The detailed results are given in Table C.22 in Appendix C.3.

For the modifications of agent capability variation and agent velocity variation, the results on the Levenshtein distances are given in Table 5.19. By definition of the INI reoptimization heuristic (see Algorithm 3), the Levenshtein distances are equal to 0 independently of the problem specification. For the modification of agent capability variation, the average Levenshtein distances of the solutions generated by the GA-based reoptimization framework remain equal to 0 for all problem specifications except for the problem instances containing one precedence and one synchronization constraint, for which an average Levenshtein distance of 0.18 results. The globally optimal solutions resulting from the BnP approach also have on average low Levenshtein distances to the initial solution of 0.12 to 0.54 for the problem instances containing some temporal constraints and an average Levenshtein distance of 2.28 for the problem instances without any temporal constraints.

In contrast to this, the solutions generated by the GA-based reoptimization framework for the problem instances corresponding to the modification of agent velocity variation range from 0.62 to 1.56 with standard deviations around 2.5 to 4.3. Compared to the modification of agent capability variation, also the Levenshtein distances of the optimal solutions generated by the BnP approach are higher. The mean values for the different problem specifications range from 5.48 to 7.56. However, the Levenshtein distances of the solutions of the conventional, randomly initialized GA are similar for both modifications with the mean values ranging from 7.52 to 10.46.

Table 5.19: Evaluation scenario 3: Results on the Levenshtein distance (mean and standard deviation (SD)) for the modifications of agent capability variation and agent velocity variation.

Mod.	Constraints in $\mathcal{I}$ sync.	prec.	INI heuristic mean	SD	GA-based reoptimization mean	SD	Conventional GA mean	SD	Branch-and-price mean	SD
Agent capability variation	0	0	0.0	0.0	0.0	0.0	9.5	2.851	2.28	3.25
	0	1	0.0	0.0	0.0	0.0	9.1	3.008	0.4	1.327
	0	2	0.0	0.0	0.0	0.0	10.46	2.934	0.12	0.475
	1	0	0.0	0.0	0.0	0.0	7.74	2.488	0.4	1.039
	2	0	0.0	0.0	0.0	0.0	7.8	2.209	0.52	1.64
	1	1	0.0	0.0	0.18	1.26	8.04	2.514	0.54	1.711
Agent velocity variation	0	0	0.0	0.0	0.9	3.151	10.2	2.173	7.56	6.658
	0	1	0.0	0.0	1.56	4.258	9.84	2.859	7.14	7.082
	0	2	0.0	0.0	0.62	2.473	9.64	3.399	7.24	6.956
	1	0	0.0	0.0	1.12	3.392	8.68	2.549	6.56	5.08
	2	0	0.0	0.0	1.08	2.576	7.52	2.67	5.58	4.06
	1	1	0.0	0.0	0.84	2.935	8.32	2.063	5.48	4.801

For the modifications of task position variation and task duration variation, the results regarding the Levenshtein distances are similar to the results obtained within the previous evaluation scenario considering different heterogeneity levels (see Section 5.3.1). For the modification of task position variation, the average Levenshtein distances of the DIH results vary between 1.0 and 1.32, the ones of the optimal solutions of the BnP approach vary between 3.54 and 4.44 with no clear influence of the number or type of temporal constraints considered. In comparison to these values, the average Levenshtein distances resulting for the modification of task duration variation are smaller and vary between 0.04 and 0.8 for the DIH reoptimization heuristic and between 0.0 and 0.64 for the BnP approach. For the DIH applied to task duration variation problem instances, the number of synchronization constraints considered within the problem instances seems to slightly increase the resulting Levenshtein distances, while the number of precedence constraints considered has no clear influence. The detailed results on the Levenshtein distances for these modifications are given in Table C.23 in Appendix C.3.

The average Levenshtein distances and their standard deviations for the results of the different solution approaches applied to the problem instances corresponding to the modifications of precedence and synchronization constraint insertion are given in Table 5.20. For the modification of precedence constraint insertion, the average Levenshtein distances of the eDIH results vary between 0.84 and 1.20 with no clear influence of the number and type of temporal constraints contained within the initial problem instance $\mathcal{I}$. These values increase slightly to a Levenshtein distance of on average 0.88 to 1.36 for the results of the GA-based reoptimization framework. The average Levenshtein distances of the solutions generated by the eDIH for the problem instances corresponding to the modification of synchronization constraint insertion are a little

Table 5.20: Evaluation scenario 3: Results on the Levenshtein distance (mean and standard deviation (SD)) for the modifications of precedence constraint insertion and synchronization constraint insertion.

Mod.	Constraints in $\mathcal{I}$ sync.	prec.	eDIH heuristic mean	SD	GA-based reoptimization mean	SD	Conventional GA mean	SD	Branch-and-price mean	SD
Precedence constr. insertion	0	0	1.16	0.987	1.16	0.987	10.54	2.685	4.38	4.024
	0	1	0.92	0.997	1.26	2.143	10.26	2.999	3.76	5.339
	0	2	1.08	0.997	1.08	0.997	10.44	2.988	4.68	5.931
	1	0	1.2	0.98	1.36	1.572	8.52	2.532	3.28	3.904
	2	0	0.84	0.987	1.1	1.652	8.12	2.188	2.56	3.401
	1	1	0.88	0.993	0.88	0.993	8.12	2.312	2.56	3.054
Synchronization constr. insertion	0	0	1.84	0.543	5.4	4.927	9.88	2.372	8.14	3.784
	0	1	1.92	0.392	3.78	3.874	9.88	2.597	8.04	3.72
	0	2	1.88	0.475	4.54	4.704	9.9	2.563	8.0	3.561
	1	0	1.8	0.6	2.62	2.331	8.14	1.822	6.42	3.269
	2	0	1.88	0.475	2.98	2.565	7.66	2.036	5.48	2.67
	1	1	1.92	0.392	2.34	1.762	7.78	1.847	5.88	3.037

higher. They vary between 1.8 and 1.92 for the different problem specifications. For this modification, also no clear influence of the number or type of constraints within the initial problem instance can be observed. By the additional application of the GA-based reoptimization framework, the average Levenshtein distances increase to 2.34 to 5.4 for the different problem specifications. The optimal solutions generated by the BnP approach on average have a higher Levenshtein distance to the initial solutions for the modification of synchronization constraint insertion, for which the average Levenshtein distances range from 5.48 to 8.14, compared to the modification of precedence constraint insertion, for which it ranges from 2.56 to 4.68. For both modifications, the Levenshtein distances of the BnP solutions tend to be lower, if synchronization constraints are contained within the initial problem instances $\mathcal{I}$. The same holds for the results generated by the conventional, randomly initialized GA, for which the absolute values of the average Levenshtein distances of the corresponding solutions are similar for both modifications ranging from 7.66 to 10.54. The results on the Levenshtein distances for the modifications of precedence and synchronization constraint deletion are given in Table C.24 in Appendix C.3. For the modification of precedence constraint deletion, the Levenshtein distances of the GA-based reoptimization framework remain equal to 0.0, i. e. the Levenshtein distances of the INI solutions, for all problems considered. For the modification of synchronization constraint deletion, average Levenshtein distances of 2.38 for the initial problem instances with two synchronization constraints and of 0.66 for the initial problem instances with one precedence and one synchronization constraint result.

A discussion of the presented results obtained for the evaluation scenario analyzing the influence of different types and numbers of temporal constraints is given in the following section.

5.4.2 Discussion

Within the previously presented evaluation scenario, the influence of precedence and synchronization constraints on the approximation ratios, calculation times and Levenshtein distances is investigated for the different solution approaches.

Overall, both precedence and synchronization constraints influence especially the solution quality, i. e. approximation ratios, and responsiveness, i. e. calculation time, of some solution approaches. However, synchronization constraints seem to have a considerably higher impact on the corresponding results than precedence constraints.

The comparison of the DIH and INI reoptimization heuristics for the modifications of task position variation and task duration variation yields interesting differences both between these modifications and between instances with and without synchronization constraints. For the modification of task position variation, the DIH reoptimization heuristic in general yields better approximation ratios than the INI approach for more than half of the investigated problem instances. However, these differences become smaller and can also become slightly negative when synchronization constraints are considered within the problem instances. In contrast to this, for the modification of task duration variation, the application of the DIH reoptimization heuristic in the conducted evaluation does not yield a better approximation ratio for any problem instance. Moreover, for the problem instances with synchronization constraints, the solutions generated by the DIH have worse approximation ratios than the INI solutions for up to 40% of the problem instances. A cause for this decrease in solution quality of the DIH solutions is the potential overapproximation of the actual insertion costs by the delete-insert heuristic (DIH, see Algorithm 5) as soon as precedence or synchronization constraints are considered within the problem instance. This overapproximation of insertion costs may cause the initial routing to be evaluated worse than another routing within the DIH, even though the actual objective function value of the solution corresponding to the initial routing is better than the modified one. This seems to happen especially often if synchronization constraints are contained within the problem instance. A reason for this might be that the potential number β_i (see Section 3.4.4) of routes affected by a temporal shift of a synchronization constrained task n_i is at least equal to two[38]. Contrary to this, precedence constraints that are related to task n_i only increase its value of β_i to be greater than 1, if task n_i is the preceding task and the subsequent task is contained within the route of another agent. Thus, the chance of overapproximating the actual insertion costs is higher, if synchronization constraints are considered within the problem instance. Consequently, for the modification of task duration variation as performed within this evaluation, the application of the INI reoptimization heuristic can be recommended over the DIH reoptimization heuristic when temporal constraints and especially synchronization constraints are considered within the problem instance. For these problem instances, the INI reoptimization approach tends to yield solutions of higher quality while at the same time having the optimal so-

[38] If a task n_i is synchronized with $l \in \mathbb{N}$ other tasks, β_i equals at least $l + 1$.

lution stability (i. e. Levenshtein distance of 0) and requiring slightly less computation time.

The solution quality resulting from the application of the extended delete-insert heuristic (eDIH, see Definition 6) to solve problem instances with an additional temporal constraint differs depending on which kind of temporal constraint is inserted to the problem instance. The results of inserting a precedence constraint into problem instances without synchronization constraints have approximation ratios that are on average 12.6% to 14.3% better than when a synchronization constraint is inserted to the same problem instances. If the initial problem instances already consider synchronization constraints, the relative difference in approximation ratio ranges between 6.2% and 7.5%. One reason for this is the same as explained above. When inserting a synchronization constraint, the chances of overapproximating the resulting waiting times and thus the overall insertion costs resulting from reinserting the tasks affected by the new temporal constraint is higher than when inserting a precedence constraint. The same holds if synchronization constraints are contained within the initial problem instance since these also increase the chance of even higher overapproximations of the actual insertion costs. Consequently, the information basis of the eDIH on which the decision on the final routing is made is more prone to include overapproximated estimates on the resulting objective function if synchronization constraints are contained within the modified problem instance compared to problem instances only containing precedence constraints. However, despite this biased information basis, the average approximation ratios obtained within this evaluation are all below a value of 1.2 as desired.

One of the consequences of the comparably high approximation ratios generated by the eDIH for the modification of synchronization constraint insertion is, that for all problem specifications on average an improvement of up to 4.7% of the approximation ratios is generated by the additional application of the GA-based reoptimization framework. This comes at the cost of increased calculation times of about 9 s to 10 s and a slightly reduced solution stability. However, if a high priority is set on the solution quality compared to the criteria of responsiveness and solution stability, the additional application of the GA-based reoptimization framework can be recommended for the modification of synchronization constraint insertion.

Another interesting effect of the temporal constraints is the influence they have on the calculation times. While there is no obvious influence on the calculation times of the different reoptimization heuristics observable, the calculation times required by both GA-based approaches within this evaluation are very similar and reduce by about 35% to 45% as soon as any temporal constraint is considered within the problem instance compared to temporally unconstrained problem instances. The results indicate a slight further decrease in calculation time the more temporal constraints are contained within the problem instance with this effect being slightly higher for synchronization than for precedence constraints. This is probably caused by much more of the routings evaluated within the evolutionary loop of the GA being found to be infeasible if temporal constraints are to be considered. For their evaluation, no timing information according to (3.5) must be determined (see Fitness evaluation in Section 4.3), which decreases the required calculation time.

In contrast to this, the calculation time required by the exact BnP approach increases with the number ob temporal constraints contained within the problem instances. This effect is especially distinct for synchronization constraints, which cause a significant increase in calculation time. This observation within this evaluation is compatible with the results obtained by Korsah [Kor11], who presented the corresponding BnP approach. Thus, the exact BnP approach is unsuitable for application within interactive MRTA optimization systems for problem instances of the investigated size (8 tasks, 3 agents) especially if synchronization constraints are contained within the problem instances since here, on average calculation times of more than 6 min for problem instances with one synchronization constraint are required, which increase to 106 min for problem instances with three synchronization constraints. Thus, the requirement of responsiveness (see [HGQ+12, DS15]) is clearly not fulfilled by the BnP approach for problems of the considered size containing synchronization constraints. In contrast to this, the results of the conducted evaluation clearly indicate the proposed reoptimization heuristics to fulfill the requirement of being responsive, independently of the number of temporal constraints considered within the problem instances.

Concluding remarks on the results obtained within the conducted evaluation are given in the following section.

5.5 Concluding Remarks on the Evaluation Results

This section summarizes the evaluation results presented in the previous Sections 5.2 to 5.4 and provides a concluding assessment regarding the suitability of the analyzed solution approaches for application in interactive MRTA optimization systems. As defined in Section 1.1, the requirements of solution quality, responsiveness of the solution approach and solution stability are important criteria for the application of solution approaches within interactive MRTA optimization systems. The obtained evaluation results indicate significant differences between the considered solution approaches w. r. t. these criteria.

Both categories of introduced reoptimization approaches, i. e. the modification-specific reoptimization heuristics and the GA-based reoptimization framework, yield very promising results w. r. t. solution quality and stability for all modifications and evaluation scenarios considered. In comparison to the reoptimization approaches, the considered optimization approaches both show major drawbacks that limit their suitability for application in interactive MRTA optimization systems:

While the exact BnP approach by definition yields optimal results w. r. t. solution quality, the stability of the solutions generated is in general slightly lower than the solution stability of both reoptimization approaches. Furthermore, the calculation time required by the BnP approach increases exponentially with the number of tasks contained within the problem instance (see Section 5.2.1). This causes the BnP approach to not satisfy the requirement on the system response time being at most 10 s in order to keep the user's attention [DS15] for medium size and larger MRTA problem instances. The calculation time required by the BnP approach furthermore is highly influenced by temporal constraints, especially synchronization constraints, contained within the problem instance (see Section 5.4.1). Consequently, the BnP approach only fulfills the requirement of solution approach responsiveness for problem instances of very limited size. The number of precedence and especially of synchronization constraints containted within the problem instance furthermore decrease the manageable size of the problem instances (see Section 5.4.2).

Compared to the BnP approach, the calculation times required by the conventional, randomly initialized genetic algorithm (GA) are more stable w. r. t. problem size (see Section 5.2.1). The consideration of temporal constraints within a problem instance even reduces the calculation time required by the GA (see Sections 5.4.1 and 5.4.2). However, for the problem size of 8 tasks and 3 agents considered in the majority of problem instances within this evaluation, the calculation time of the GA does in general not meet the threshold of at most 10 s system response time (see Sections 5.2.1 and 5.3.1) in order to keep the user focused on the interaction [DS15]. Only under the consideration of temporal constraints, this threshold can be met for the majority of problem instances (see Sections 5.4.1). What is more, within the conducted evaluation, the conventional, randomly initialized GA clearly showed the worst performance of all evaluted solution approaches w. r. t. solution quality and solution stability (see Sections 5.2.1, 5.3.1 and 5.4.1). Since the extension of the conventional, randomly initialized GA towards the GA-based reoptimization framework substantially improves the generated solutions w. r. t. these two criteria while only marginally influencing the required calculation time, the GA-based reoptimization framework can be clearly prioritized over the conventional GA for application in interactive MRTA optimization systems. Consequently, both proposed reoptimization approaches, the modification-specific reoptimization heuristics and the GA-based reoptimization framework, clearly outperform the evaluated optimization approaches w. r. t. the combination of criteria relevant for their application in interactive MRTA optimization systems.

Between the two reoptimization approaches, there is a big difference w. r. t. their responsiveness, evaluated based on their required calculation time. Out of all solution approaches under evaluation, only the proposed reoptimization heuristics yield calculation times that for all modifications and evaluation settings considered are by orders of magnitude below 300 ms and thus suitable for instantaneous, closed-loop interactions [DS15]. Even though there seems to be a linear increase in required calculation time with the problem size, the calculation times obtained within this evaluation of all reoptimization heuristics are all more than factor 10 below the threshold of 300 ms, such that the reoptimization heuristics can be expected to meet the responsiveness requirements for instantaneous interactions up to comparably large problem instances. Furthermore, the reoptimization heuristics also have the potential to handle multiple modifications at a time (by the consecutive application of several reoptimization heuristics) in a responsive manner suitable for instantaneous, closed-loop interaction.

Compared to this, the GA-based reoptimization framework requires by orders of magnitude higher calculation times, which are very similar to the calculation times required by the common, rondomly intitialized GA. Thus, the threshold of 300 ms for instantaneous interactions can clearly not be met and also the threshold of 10 s in order to maintain the user's attention is in general exceeded for the unconstrained problem instances considered within the conducted evaluation.

While it is important to note, that the implementations of all solution approaches within this thesis were not fully optimized w. r. t. calculation times, the obtained results still give a good indication of the orders of magnitudes of calculation times required, the influences diverse problem features have on the calculation times, and the relation between the calculation times required by the different solution approaches under consideration. However, there probably is potential for further improvement w. r. t. calculation time. This holds especially for the GA-based solution approaches, for which the calculation time is strongly influenced by the numbers of evolutionary loops conducted, such that different termination criteria might be considered in order to yield a better responsiveness. The criteria of a fixed number of 100 iterations chosen within this thesis however allows for an objective comparison between the conventional, randomly intitialized GA and the GA-based reoptimization approach w. r. t. all criteria relevant for interative MRTA reoptimization settings.

Comparing the two reoptimization approaches w. r. t. the other two criteria relevant for interactive MRTA optimization systems, the results of this evaluation indicate that in the majority of temporally unconstrained problem instances, the solution quality of the solution generated by the respective reoptimization heuristic is not or only marginally improved by the GA-based reoptimization framework. One exception from this is the modification of synchronization constraint insertion, for which the approximation ratio, i. e. the criterium evaluating the solution quality, was improved by up to on average 4.7% for synchronization constrained problem instances (see Section 5.4.2). However, the quality of the solutions resulting for all modifications are all close to the optima with approximation ratios $\alpha < 1.2$. Thus, also for the modifications of task insertion and task deletion, the problem instances considered within this evaluation all yield solutions with objective function values that are noticeably better than the guarantees on the tight upper bounds of the resulting approximation ratios (see Sections 3.2.5, 5.2.1, 5.3.1 and 5.4.1). Moreover, the obtained low approximation ratios indicate, that also if multiple reoptimization heuristics were applied consecutively, the resulting solution will in many cases still be close to optimal.

Furthermore, by their definition all reoptimization heuristics yield stable solutions with acceptable Levenshtein distances ranging from zero to at most two. For the problem instances in which the GA-based reoptimization framework alters the solutions of the reoptimization heuristics, the solution stability in general slightly decreases.

Overall, the reoptimization heuristics outperform the GA-based reoptimization framework w. r. t. responsiveness and also perform better w. r. t. solution stability (see Sections 5.2.1, 5.3.1 and 5.4.1). In return, the GA-based reoptimization framework on average only improves the solution quality by less than 2% for most modifications. Only for the modification of synchronization constraint insertion an higher average improvements of up to 4.7% results (see Sections 5.2.1, 5.3.1 and 5.4.1). Therefore, the reoptimization heuristics can be considered to provide the best combination to fulfill the criteria for application in interactive MRTA optimization systems, i. e. solution quality, solution stability and responsiveness, and to be very well suited for this kind of application. However, if a high priority is set on solution quality, the additional application of the GA-based reoptimization framework can be advisable. This holds especially for the modification of synchronization constraint insertion when applied to temporally constrained problem instances (see Section 5.4.2).

6 Conclusion

This thesis introduces, analyzes and evaluates for the first time centralized reoptimization approaches for heterogeneous, time-extended MRTA problems with precedence and synchronization constraints, aiming at enabling interactive MRTA optimization systems.

Interactive MRTA optimization systems allow for combining the strengths of an automated system in terms of calculation performance with the strengths of the human in situational awareness and capability of abstraction. This is realized by letting a human define MRTA problems and adapt it to dynamic environmental changes, while the automated system solves the defined problem instances. To enable such interactive systems, solution approaches that unite guaranteed solution quality, responsiveness, and solution stability are necessary. Solution quality is essential due to its influence on MRSs' performance. Responsive systems, that in the best case allow for instantaneous, closed-loop user interaction, together with solution stability, are relevant criteria for user acceptance and satisfaction. However, the analysis of the current state of research reveals that there is no adequate solution approach in literature for heterogeneous, time-extended MRTA problems with precedence and synchronization constraints that combines these requirements.

To close this gap, the first contribution of this thesis is the introduction of reoptimization heuristics for ten modifications of heterogeneous, time-extended MRTA problems with precedence and synchronization constraints. These reoptimization heuristics are the first for the considered problem class for which guarantees on the resulting solution quality are derived.

For all modifications, the solutions generated by the respective reoptimization heuristics are guaranteed to be feasible and thus to fulfill all constraints. For eight out of the ten modifications, it is moreover guaranteed that, if the solution space is non-empty, a feasible solution is found by the proposed reoptimization heuristics. Even stronger performance guarantees in the form of upper bounds of $\alpha \leq 2$ on the resulting approximation ratios α are given for the two modifications of inserting and deleting a task. These modifications are the most studied ones also for the TSP (see Table 2.4) and belong to the most important ones in practical MRTA applications, especially the insertion of a task (see Table 2.3). The bounds given on the approximation ratios for these modifications are proven to hold for all reoptimization approaches that fulfill two conditions. For the modification of task insertion, it is shown that the proposed CMI reoptimization heuristic always fulfills these conditions and thus yields an approximation ratio bounded above by $\alpha \leq 2$, which is furthermore proven to be a tight bound. For problem instances not considering any precedence or synchronization constraints, even smaller tight upper bounds on the approximation ratio down to $\alpha \leq 3/2$ are derived. Also, the TDH reoptimization heuristic proposed for the modification of task deletion is shown to fulfill the respective conditions in the vast majority of instances for which the resulting approximation ratios are thus bounded above by $\alpha \leq 2$. This bound is also shown to be tight. Furthermore, a tight upper bound of $\alpha \leq 3/2$ is derived for problem instances with homogeneous agents and without any precedence or synchronization constraints.

The second contribution of this thesis is the introduction of a metaheuristic reoptimization framework. This framework combines the previously defined reoptimization heuristics with metaheuristic solution approaches that aim at balancing intensification and diversification. Any metaheuristic suitable for heterogeneous, time-extended MRTA problems with precedence and synchronization constraints can be applied. It is shown that for any metaheuristic that guarantees to retain the best solution generated throughout the search process, the performance guarantees given for the respective reoptimization heuristics remain valid. Moreover, a specific realization of the metaheuristic reoptimization framework is proposed. It is based on a GA and fulfills the aforementioned property of preserving the performance guarantees given for the respective reoptimization heuristics.

To assess the performance of the proposed reoptimization approaches on heterogeneous, time-extended MRTA problems instances with precedence and synchronization constraints, an extensive evaluation of the reoptimization heuristics and the GA-based reoptimization framework is conducted. The reoptimization approaches are compared to two optimization approaches, an exact BnP approach and a conventional GA. The influence of different features of the problem instances including problem size, heterogeneity and the number of precedence and synchronization constraints considered, is analyzed. The results reveal that both reoptimization approaches outperform the optimization approaches w. r. t. the criteria relevant for application in interactive MRTA optimization systems. Especially the reoptimization heuristics by far yield the best combination of solution quality, solution stability, and responsiveness of the solution approach. For all problem specifications evaluated, the reoptimization heuristics yield solutions with close to optimal quality and with a deterministically high solution stability in a very responsive manner that allows for instantaneous, closed-loop interactions. Only if high priority is set on the solution quality, an additional application of the GA-based reoptimization framework can be recommended, especially if synchronization constraints are contained in the problem instance.

Both, the theoretical performance guarantees and the evaluation results confirm the introduced reoptimization heuristics and the metaheuristic reoptimization framework to be very well suited for application in interactive MRTA optimization systems. Whenever modifications to heterogeneous, time-extended MRTA problems with potential precedence and synchronization constraints occur, they generate stable, high quality solutions in a very responsive manner. This allows for an increase in user acceptance and MRS performance and thereby renders the reoptimization approaches a key enabler for a beneficial application of MRSs in volatile and unstructured environments such as space exploration, inspection of power plants, and nursing practice. Thus, this thesis provides an important step towards broader purposeful application of MRSs for society and eventually towards the anticipated improvement of the standard of living.

A Fundamentals Regarding the Modeling of MRTA Optimization Problems

A.1 Mathematical Models for the MTSP Optimization Problem

The basics of two common approaches to model MTSPs and their relaxations are given in this section. They comprise a three-index model based on a vehicle-flow formulation[39] and a two-index model based on a set partitioning formulation. To focus on the fundamentals of the formulations, the following depictions consider the basic MTSP problem, i. e. additional constraints are neglected and a homogeneous team of agents is considered. However, both formulations allow for adaptions in order to represent problem features of time-extended MRTA problems (or VRPs).

A.1.1 Three-Index Model Based on a Vehicle-Flow Formulation

The three-index vehicle-flow model, which was initially proposed by Golden et al. [GMN77], is presented in the following. The given exposition is based on [BBV08], but has been shortened and adapted to focus on the basic MTSP. Vehicle-flow formulations are especially well suited for cases in which the objective function depends on costs associated with agents traveling between distinct positions and on agents being allocated to tasks [TV02].

A complete graph $G(V,E)$ is used to model the problem. Let the set of agents be denoted by $\mathcal{K} = \{k_1, k_2, \ldots, k_K\}$ and the set of tasks be given as $\mathcal{N} = \{n_1, n_2 \ldots, n_N\}$. The vertices of the graph comprise all tasks plus the agents' common depot which is denoted as $o_{\mathcal{K}}$, thus $V = \mathcal{N} \cup \{o_{\mathcal{K}}\}$. Binary decision variables $x_{i,j}^m$ are equal to one if edge $(i,j) \in E$ is traversed by agent $m \in \mathcal{K}$ and zero otherwise. With $\delta(i,j)$ denoting the cost of traversing edge $(i,j) \in E$, the tree-index vehicle-flow model is given as

$$\min \sum_{m \in \mathcal{K}} \sum_{\substack{i,j \in V \\ i \neq j}} \delta(i,j) x_{i,j}^m \tag{A.1}$$

[39] Besides the three-index vehicle-flow model, also a two-index vehicle-flow formulation exists. However, the two-index variants do not explicitly differentiate between the individual vehicles. Thus, they do not allow for a direct allocation of agents to tasks which is also why these models are not suited for problem instances in which route costs depend on the allocated agent. This important feature can, in contrary, be modeled by the three-index variation of the vehicle-flow formulation. [TV02]
Therefore, only the three-index model is presented.

$$\text{s.t.} \sum_{m \in \mathcal{K}} \sum_{i \in V} x^m_{i,j} = 1 \qquad \forall j \in V \setminus \{o_{\mathcal{K}}\} \tag{A.2a}$$

$$\sum_{m \in \mathcal{K}} \sum_{i \in V} x^m_{i,l} - \sum_{m \in \mathcal{K}} \sum_{j \in V} x^m_{l,j} = 0 \qquad \forall l \in V \setminus \{o_{\mathcal{K}}\}, \forall m \in \mathcal{K} \tag{A.2b}$$

$$x^m_{i,j} \in \{0,1\} \qquad \forall i,j \in V, i \neq j, \forall m \in \mathcal{K}. \tag{A.2c}$$

In the above formulation, the optimization objective (A.1) is to minimize the overall costs. Constraint (A.2a) enforces all tasks to be visited exactly once and constraint (A.2b) makes sure that every agent departs from all tasks it visits. The binarity of the decision variables are ensured by constraint (A.2c).

Vehicle flow formulations are very commonly used to model VRPs and allow for the consideration of various constraints (see e. g. [BR07, KLB09, DRL11, RJDL12, LIW$^+$23]). On the other hand, vehicle flow models can have very weak linear programming relaxations when constraints are tight[40]. [TV02]

A.1.2 Two-Index Model Based on a Set Partitioning Formulation

A main advantage of the set partitioning formulation is that its linear programming relaxation is typically much tighter than the one of the vehicle-flow formulation [TV02]. The set partitioning formulation was initially proposed by Balinski and Quandt [BQ64]. The explanations given in the following are based on [TV02, Chapter 1.3.4].

The set partitioning formulation is based on the set of feasible routes r starting and ending at the depot. Routes are circuits of the graph $G(V,E)$ which can be defined as sequences of edges or vertices. The set of feasible routes is given by $\mathcal{R}$. Each route is associated with a cost κ_r. The binary indicator variable b_{ir} takes the value 1 if vertex $i \in V$ is covered by route $r \in \mathcal{R}$ and 0 otherwise. Binary decision variables λ_r are equal to 1 if and only if route $r \in \mathcal{R}$ is chosen. Using this notation, set partitioning formulation is given by

$$\min \sum_{r \in \mathcal{R}} \kappa_r \lambda_r \tag{A.3}$$

$$\text{s.t.} \sum_{r \in \mathcal{R}} b_{ir} \lambda_r = 1, \qquad \forall i \in \mathcal{N} \tag{A.4a}$$

$$\lambda_r \in \{0,1\} \qquad \forall r \in \mathcal{R}. \tag{A.4b}$$

The cost of the selected routes is minimized by the optimization objective (A.3). The set partitioning constraints (A.4a) impose that each task is visited exactly once by the routes selected and (A.4b) ensures the binarity of the decision variables.

40 An inequality constraint is tight in a point x if it is fulfilled with equality for x [PAN23, p. 126].

The set partitioning formulation allows for very general route costs and can be extended to express various constraints (see e. g. [HRJL08, Kor11, KKB+12]). A main advantage of the formulation is that its linear programming relaxation is typically much tighter than the one of the vehicle-flow formulation. On the downside however, set partitioning models come at the cost of a very large number of variables. [TV02]

Both, the three-index vehicle flow and the two-index set partitioning formulation, are often applied in optimization-based solution approaches. Both models can be extended to express various problem variants.

A.2 Linearization of the Vehicle Flow Formulation of the MRTA Optimization Problem

The MRTA optimization problem representation given Problem 3.1 is non-linear due to constraint (3.2f) and due to the representation of waiting costs $c^w_{\mathcal{I}}$ in the objective function (3.1). This section introduces the linearization of the aforementioned components and thus of the MRTA optimization problem.

For this linearization, additional decision variables $w_{i,j} \in \mathbb{R}_{\geq 0}$ representing the waiting times between two vertices $i \in V$, $j \in \mathcal{N}$ are introduced.[41] Thus, the waiting times associated with a solution $c^w_{\mathcal{I}}(\vec{X}_{\mathcal{I}})$ can be linearly expressed as

$$c^w_{\mathcal{I}}(\vec{X}_{\mathcal{I}}) = \sum_{i \in V} \sum_{j \in \mathcal{N}} w_{i,j}. \tag{A.5}$$

Using a sufficiently large constant $T_{\max} \in \mathbb{R}_{\geq 0}$, the appropriate determination of waiting times can be ensured by the additional constraints

$$w_{i,j} \geq t_j - t_i - \sum_{m \in \mathcal{K}} x^m_{i,j} \left(d^m_i + d^m_{i,j}\right) - \left(1 - \sum_{m \in \mathcal{K}} x^m_{i,j}\right) T_{\max}, \quad \forall\, i \in \mathcal{N}, j \in \mathcal{N} \tag{A.6a}$$

$$w_{i,j} \geq t_j - t_i - \sum_{m \in \mathcal{K}} x^m_{i,j} d^m_{i,j} - \left(1 - \sum_{m \in \mathcal{K}} x^m_{i,j}\right) T_{\max}, \quad \forall\, i \in \mathcal{O}, j \in \mathcal{N} \tag{A.6b}$$

$$w_{i,j} \in \mathbb{R}_{\geq 0} \quad \forall\, i \in V, j \in \mathcal{N}, \tag{A.6c}$$

which must be added to the optimization problem. In case an agent $m \in \mathcal{K}$ transitions from a task $i \in \mathcal{N}$ to a task $i \in \mathcal{N}$, i. e. $\sum_{m \in \mathcal{K}} x^m_{i,j} = 1$, constraint (A.6a) ensures that the waiting time between the two tasks cannot become smaller than the difference between their starting times minus the sum of the agent-dependent duration d^m_i of task i and the transition time $d^m_{i,j}$ to task j. In case edge (i, j) is not taken by any agent, i. e. $\sum_{m \in \mathcal{K}} x^m_{i,j} = 0$, a sufficiently large choice of $T_{\max} \in \mathbb{R}_{\geq 0}$ ensures $w_{i,j}$ to be bounded below by a negative value which becomes negligible due to constraint (A.6c).

[41] Since waiting times, which are caused by precedence or synchronization constraints, can only occur in front of tasks, only $j \in \mathcal{N}$ must be considered.

The same holds for transitioning from a depot to another vertex, as given by constraint (A.6b). The only difference in this case is that a depot is not associated with an agent-dependent duration and thus d_i^m can be neglected. To complete the linearization, the same technique can be applied to replace the non-linear constraint (3.2f), which ensures the consistency of starting times, by the linearized constraint

$$t_j - t_i - \sum_{m \in \mathcal{K}} x_{i,j}^m \left(d_i^m + d_{i,j}^m \right) \geq \left(\sum_{m \in \mathcal{K}} x_{i,j}^m - 1 \right) T_{\max}, \qquad \forall i \in V, j \in \mathcal{N}. \tag{A.7}$$

An appropriate choice of $T_{\max}$ allows for the application of the same constant $T_{\max}$ in (A.6) and (A.7).

B Branch-and-Price for Time-Extended MRTA Problems with Complex Dependencies

This chapter is concerned with exact solution approaches for time-extended MRTA problems. As revealed by the state of the research presented in Section B.1, so far no exact solution approach exists that is capable of handling complex dependencies. Thus, the exact *branch-and-price (BnP)* approach applied within this thesis to generate globally optimal solutions for time-extended MRTA problems with cross-schedule dependencies, which is briefly introduced in Section B.2, is extended towards the consideration of complex dependencies. Three different approaches to this problem are presented in Section B.3 and a comparative evaluation of these approaches is given in Section B.4.

B.1 Exact Optimization-Based Solution Approaches

Contrary to heuristic or metaheuristic approaches, exact approaches rely on rigorous algorithms to achieve globally optimal solutions. In the following, an overview of existing exact solution approaches for time-extended MRTA problems is given. The number of exact solution methods introduced in the MRTA problem domain is very limited, which is why also related problem domains are taken into consideration. Since this thesis focuses on time-extended MRTA problems that solve both task allocation and task scheduling (see Section 2.2), which are relevant for many practical applications, only approaches for problems with cross-schedule or complex dependencies are considered. The presented approaches are investigated on whether they take into consideration the required MRTA problem features as defined in Section 1.2, i. e. heterogeneous tasks and agents, precedence and synchronization constraints.

Bredström and Rönnquvist [BR07] propose a BnP[42] algorithm for a VRP. They consider homogeneous agents, time windows for customer service as well as for agent availability and synchronization constraints between tasks, i. e. customers. Each synchronization constraint cannot involve more than two customers. The optimization objective is to minimize traveling time and the sum of arbitrarily set weights for each agent serving each destination.

A HCCSP is solved by Hansen et al. [HRJL08]. Agents, i. e. caretakers, with different competences have to be allocated and scheduled to tasks, i. e. customers, that require

[42] Branch-and-price is an exact solution approach for (mixed-)integer linear optimization problems. It combines branch-and-bound and column generation methods. [DL11]

different services. Some services require the presence of several caretakers and thus impose synchronization constraints on the problem. Furthermore, service time windows and caretakers working hours must be respected. The objective function considers multiple criteria such as the minimization of overall operation cost and the maximization of the level of service, which is e. g. influenced by the number and priority of services that cannot be scheduled. The problem is solved using a BnP approach.

A BnP approach applied to time-extended MRTA problems comprising a comprehensive amount of different types of constraints is proposed by Korsah [Kor11]. She considers temporal constraints including precedence, synchronization and time windows, but also location choice, capacity constraints and location proximity constraints. The considered optimization objective includes rewards for each task performed as well as costs associated with travel and waiting times. A BnP method is proposed to solve the problem.

Dohn et al. [DRL11] consider a VRP with time windows. Precedence and synchronization constraints are modeled using generalized precedence constraints which also allow for the representation of temporal overlapping constraints and constraints on minimum or maximum temporal difference in starting times of task execution. They propose four different problem formulations and apply a branch-and-cut-and-price (BCnP)[43] approach to all of them.

A BnP approach for the HCCSP is introduced by Rasmussen et al. [RJDL12]. They consider the agents, i. e. caretakers, to have heterogeneous capabilities and tasks, i. e. visits, to be constrained by means of generalized precedence constraints and time windows. The objective is given as a weighted sum of minimizing the amount of tasks not dealt with, maximizing rewards associated with each agent-task allocation, and minimizing total travel costs. Rasmussen et al. [RJDL12] propose a clustering of tasks in order to accelerate calculation time. Since this restricts the solution space, solution optimality can no longer be guaranteed if task clustering is applied.

Haddadene et al. [HLP14] consider also a HCCSP. They take agents as homogeneous w. r. t. to travel times, but assume each agent to be only able to perform one out of several types of tasks. Precedence and synchronization constraints as well as time windows must be respected. They propose two different problem formulations which differ in whether the objective function additionally to agent-task-allocation and travel dependent costs also considers waiting times. They solve both models using CPLEX®[44].

Another HCCSP with time windows, precedence and synchronization constraints which is solved by CPLEX® is given by Taieb et al. [TLM19]. Allocation-dependent costs that depend on the individual mapping between agents and tasks as well as travel costs are to be minimized. Their model additionally takes into consideration breaks required by the caretakers, i. e. agents.

43 BCnP is an extension to BnP for solving (mixed-)integer linear optimization problems that additionally applies cutting planes to strengthen the linear programming relaxations. [DL11]

44 CPLEX® is a commercial optimizer by IBM [IBM].

Li et al. [LIW⁺23] consider a truck-drone routing problem (TDRP) with simultaneous delivery and pickup. The heterogeneous set of agents consists of trucks and drones and good delivery and pickup services for customers represent the tasks. A set of homogeneous trucks, each equipped with its own set of heterogeneous drones have to deliver a defined set of goods to customers and can furthermore fulfill any subset of a known set of pickup requests from customers to increase the objective function values. Drones can only start their routes when the truck arrives at a customer and a truck must wait at until all drones return before it can continue its drive. Thus, the positions of the depots for the drones are variable and coupled with the routed of the trucks. Solutions are evaluated by their associated service and transportation costs reduced by the utility generated by pickup services realized. The authors introduce a BCnP algorithm to optimally solve the problem.

Table B.1 summarizes the presented publications to time-extended MRTA and related problems. The majority of the respective exact solution approaches belongs to the domain of VRPs or HCCSPs, and they allow for the consideration of various different constraints. It is interesting to note that all exact optimization-based approaches presented consider time-extended MRTAs with cross-schedule dependencies. To the best of the author's knowledge, no exact optimization-based solution approaches for time-extended MRTA problems with complex dependencies exist. Thus, the consideration of task decomposition has not yet been tackled by the respective solution approaches. To close this gap, three approaches on how to consider complex dependencies in BnP are presented in Section B.3. They are based on the BnP approach by Korsah [Kor11], which is briefly introduced in the following section.

Table B.1: Overview of exact solution approaches to solve time-extended MRTA or related optimization problems. The depicted approaches all consider a time-extended planning horizon. For each publication, the applied solution approach, the problem domain as well as the degree of interrelatedness (Interr.) are given. Furthermore, it is indicated, whether heterogeneous teams of agents (Het.), precedence constraints (Prec.) and synchronization constraints (Sync.) are considered. Additionally respected constraints are given in column labeled "Others". All approaches respect time windows.

Publication	Solution approach	Problem domain	Interr.	Het.	Prec.	Sync.	Others
[BR07]	BnP	VRP	XD	✗	✗	(✓)	
[HRJL08]	BnP	HCCSP	XD	✓	✓	✓	working hours
[Kor11]	BnP	MRTA	XD	✓	✓	✓	location choice and proximity constraints
[DRL11]	BCnP	VRP	XD	✗	✓	✓	
[RJDL12]	BnP	HCCSP	XD	✓	✓	✓	
[HLP14]	CPLEX®	HCCSP	XD	(✓)	✓	✓	
[TLM19]	CPLEX®	HCCSP	XD	✓	✓	✓	breaks for agents
[LIW⁺23]	BCnP	TDRP	XD	✓	✓	✗	moving depots

B.2 Application of Branch-and-Price to Time-Extended MRTA Problems

To solve the heterogeneous, time-extended MRTA problem with precedence and synchronization constraints, i. e. an MRTA problem with cross-schedule dependencies, a BnP approach which uses a set-partitioning formulation (see Section A.1.2) is applied. BnP approaches combine branch-and-bound with column generation for solving huge integer linear programs (ILPs). An introduction to branch-and-bound and to column generation can for example be found in [ES22, Chapter 5.3.2] and [Lüb11], respectively. The approach applied in this thesis is adapted from [Kor11], since this is the only exact optimization approach in the MRTA domain that considers heterogeneous agents, precedence and synchronization constraints (see Table B.1). Thus, it considers all features of the time-extended MRTA Problem 3.1 of this thesis, i. e. heterogeneous agents w. r. t. capabilities and velocities, precedence and synchronization constraints (see also Section 1.2).[45]

The relevant sets, decision variables and constants are defined as given in Table B.2. The different time variables and constants are exemplarily illustrated in Figure B.1 and explained in the following. The constant start times t_{ir}^{m} represent the time the execution of task n_i would start on route $r \in \mathcal{R}^m$ if there was no waiting time. They are determined by the vertices contained within a route $r \in \mathcal{R}^m$ and their order. An agent k_m having to wait right before the execution of task n_i (caused by a precedence or synchronization constraint) causes a delay ζ_i^m. This delay is carried forward to all following tasks n_j within a route resulting in a delay ζ_{ij}.

[45] Further problem features considered by Korsah [Kor11] are disregarded, i. e. location choice, non-overlapping constraints and proximity constraints. Furthermore, all compund tasks are assumed to be decomposed into their simple subtasks.

Table B.2: Definition of variables applied in the BnP approach.

	Symbol	and interpretation	Domain
Relevant sets	$\mathcal{K}$	set of all agents	
	$\mathcal{N}$	set of all tasks	
	$\mathcal{R}^m$	set containing all possible routes of agent k_m	
	$\mathcal{P}$	set of precedence constraints	
	$\mathcal{S}$	set of synchronization constraints	
Decision variables	λ_r^m	whether agent k_m is allocated to route $r \in \mathcal{R}^m$	$\in \{0,1\}$
	t_i	starting time of task n_i	$\in \mathbb{R}_{\geq 0}$
	ζ_i^m	delay of agent k_m for task n_i	$\in \mathbb{R}_{\geq 0}$
	ζ_{ji}	delay of task n_i caused by task n_j	$\in \mathbb{R}_{\geq 0}$
Constants and indicator variables	b_{ir}^m	whether task n_i is on route $r \in \mathcal{R}^m$	$\in \{0,1\}$
	b_{jir}^m	whether vertex j is scheduled before vertex i on route $r \in \mathcal{R}^m$	$\in \{0,1\}$
	d_i^m	execution time needed by agent k_m for task n_i	$\in \mathbb{R}_{\geq 0}$
	$d_{i,j}^m$	transition time of agent k_m from vertex i to vertex j	$\in \mathbb{R}_{\geq 0}$
	κ_{1r}^m	costs of route $r \in \mathcal{R}^m$ of agent k_m neglecting waiting times	$\in \mathbb{R}_{\geq 0}$
	κ_2^m	costs related to waiting times of agent k_m	$\in \mathbb{R}_{\geq 0}$
	t_{ir}^m	start time of task n_i on route $r \in \mathcal{R}^m$ neglecting waiting time	$\in \mathbb{R}_{\geq 0}$
	T_{max}	maximum planning horizon	$\in \mathbb{R}_{\geq 0}$

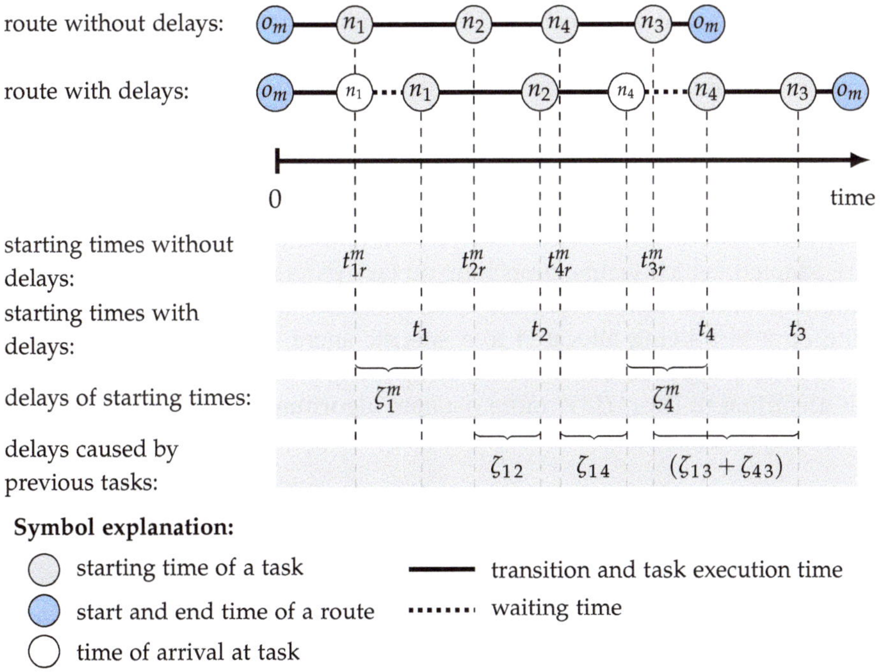

Figure B.1: Illustration of the different time variables and constants applied in the BnP approach. The figure is adapted from [Wö23].

Using these definitions, the problem formulation is given as

$$\min \left[\sum_{m\in\mathcal{K}} \sum_{r\in\mathcal{R}^m} \kappa_{1r}^m \lambda_r^m + \sum_{m\in\mathcal{K}} \sum_{i\in\mathcal{N}} \kappa_2^m \zeta_i^m \right] \tag{B.1}$$

$$\sum_{r\in\mathcal{R}^m} \lambda_r^m = 1 \qquad \forall\ m \in \mathcal{K} \tag{C1}$$

$$\sum_{m\in\mathcal{K}} \sum_{r\in\mathcal{R}^m} b_{ir}^m \lambda_r^m = 1 \qquad \forall\ i \in \mathcal{N} \tag{C2}$$

$$t_i - \sum_{m\in\mathcal{K}} \sum_{r\in\mathcal{R}^m} t_{ir}^m b_{ir}^m \lambda_r^m - \sum_{j\in\mathcal{N}} \zeta_{ji} - \sum_{m\in\mathcal{K}} \zeta_i^m = 0 \qquad \forall\ i \in \mathcal{N} \tag{C3}$$

$$-\zeta_{ji} + \sum_{m\in\mathcal{K}} \zeta_j^m + T_{max} \sum_{m\in\mathcal{K}} \sum_{r\in\mathcal{R}^m} b_{jir}^m \lambda_r^m \leq T_{max} \qquad \forall\ i \in \mathcal{N}, j \in \mathcal{N} \tag{C4a}$$

$$\zeta_{ji} - T_{max} \sum_{m\in\mathcal{K}} \sum_{r\in\mathcal{R}^m} b_{jir}^m \lambda_r^m \leq 0 \qquad \forall\ i \in \mathcal{N}, j \in \mathcal{N} \tag{C4b}$$

$$\zeta_{ji} - \sum_{m\in\mathcal{K}} \zeta_j^m \leq 0 \qquad \forall\ i \in \mathcal{N}, j \in \mathcal{N} \tag{C4c}$$

$$t_i + \sum_{m\in\mathcal{K}} \sum_{r\in\mathcal{R}^m} d_i^m b_{ir}^m \lambda_r^m - t_j \leq 0 \qquad \forall\ (i,j) \in \mathcal{P} \tag{C5}$$

$$t_i - t_j = 0 \qquad \forall\ (i,j) \in \mathcal{S}. \tag{C6}$$

As in (3.1), the optimization objective (B.1) is the minimization of the weighted sum of delay and routing costs which consider both driving as well as task processing times. The problem constraints require each agent to perform one (potentially empty) route (C1) and each task to be performed exactly once (C2). Constraints (C3)-(C4c) ensure the time and delay variables to be consistent and (C5) and (C6) model the precedence and synchronization constraints, respectively.

To exactly solve the MRTA problem, the relevant branching rules of [Kor11] and their order are adapted, which yields: branching on task pairs occurring on the same route, branching on pairwise task's order for tasks allocated to the same agent's route and branching on a task being allocated to a specific agent. To generate new profitable routes, the pricing subproblem is solved using a depth-first search and column management according to [Kor11]. A more detailed algorithmic description can be found in [Kor11, Chapter 5].

B.3 Consideration of Complex Dependencies

In this section, three approaches on how to extend the previously introduced BnP approach to handle also compex tasks and thus heterogeneous, time-extended MRTA problems with precedence and synchronization constraints and complex dependencies, are introduced. These approaches have been established in the master thesis of

Wöran [Wö23] supervised by the author and have been published in conference proceedings by the author [BWRH24]. They are based on modeling complex tasks as task trees, which is presented in the following section.

B.3.1 Modeling of Complex Tasks

In this section, the modeling of complex tasks (see Section 2.1) using task trees as introduced by Zlot [ZS06] is presented. This representation of complex tasks is applied in the following in order to extend the problem defined in Section B.2 towards complex tasks.

Task trees apply AND and OR operators to define how tasks are interconnected. Temporal constraints are allowed but only between simple tasks which both are direct subnodes of the same AND-operator. One exemplary task tree with temporal constraints is shown in Figure B.2. Within this example, the overall problem contains seven simple tasks. Tasks n_6 and n_7 must both be executed with n_7 proceeding n_6. Additionally, either task n_1, or tasks n_2 and n_3, or tasks n_4 and n_5 must be executed by the agents. If tasks n_2 and n_3 are chosen, these are related by a precedence constraint $p_{2,3} = 1$, and if tasks n_4 and n_5 are chosen, they are related by a synchronization constraint $s_{4,5} = 1$.

The consideration of complex tasks changes the MRTA problem with cross-schedule dependencies presented in Section B.2. The combination of all simple tasks (all leafs in a task tree) which originate from the complete decomposition of the original tasks from the overall problem result in the set of fully decomposed simple tasks $\mathcal{N}$. If complex tasks are contained within the problem instance, only a subset of $\mathcal{N}$ must be executed. Thus, not only must an optimal task allocation and task scheduling be determined, but also must the problem of task decomposition be solved to optimality. These problems influence each other.

To handle this extended MRTA problem, three approaches called *decomposition method, decision variable method* and *cluster method* are presented in the following.

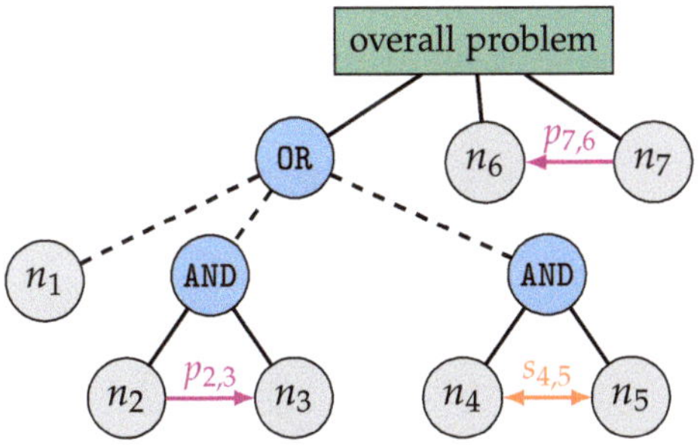

Figure B.2: Exemplary task tree with time constraints. The overall problem is located at the top and each of its subnodes must be fulfilled. Operators are labeled with "AND" or "OR". All leafs of the tree represent simple tasks. Precedence constraints and synchronization constraints are visualized with purple and orange arrows, respectively. The figure is adapted from [Wö23].

B.3.2 Decomposition Method

The decomposition method considers all possible decompositions of the complex tasks contained in a problem instance in a complete enumeration approach.

In a first step all *minimal decompositions* of the overall problem must be determined. A *decomposition* contains a subset of the simple tasks of the overall task set $\mathcal{N}$ which must be part of the solution. A decomposition is considered to be *minimal* if the decomposition solves the overall problem, but no strict subset of the tasks of the decomposition does. For the example depicted in Figure B.2 the set of tasks $\{0,5,6\}$ is a minimal decomposition because none of its strict subsets fulfils the overall problem but the set itself does. On the other hand, the set $\{0,2,5,6\}$ also fulfils the overall problem, but it is not minimal, because $\{0,5,6\}$ is a strict subset of it. All minimal decompositions and corresponding task trees only containing simple tasks of the example depicted in Figure B.2 are given in Figure B.3.

All minimal decompositions of a problem instance can be determined with an iterative approach: The problem instance itself is modeled as a task tree with the root node being an `AND` operator. Each `AND` and `OR` operator contains a set of suboperators and a set of simple subtasks, both of which can potentially be empty. Starting at the root node, the possible decompositions are determined iteratively. For each `AND` operator, its simple suptasks are added to all possible decompositions determined so far. For each `OR` operator, new decompositions are created for each subtask and for each possible decomposition of its suboperators. Starting at the root node, the decompositions are determined by iteratively assessing all subtasks and suboperators.

After the determination of all minimal decompositions of the overall problem, for each minimal decomposition a separate subproblem is created that only contains the set of simple tasks necessary for the respective decomposition. Each of these subproblems is an MRTA problem with at most cross-schedule dependencies (see Section 2.1), i. e. all simple task must be part of the solution. This reduces the complexity of the problem to only contain the problems of task allocation and task scheduling. As a result, each subproblem can be modeled and solved by the original BnP approach described in Section B.2. After the solution is calculated for every possible subproblem, the one with the smallest objective function value solves the overall problem globally optimal.

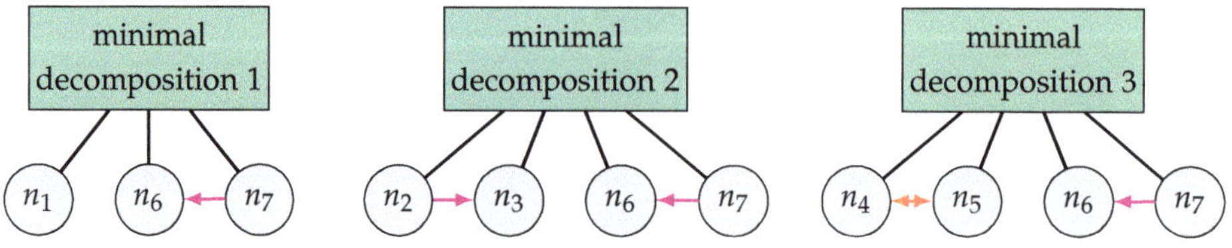

Figure B.3: Minimal decompositions of the exemplary MRTA problem with complex dependencies depicted in Figure B.2. The figure is adapted from [Wö23].

In summary, the decomposition approach separates the problem of task decomposition from the problems of task allocation and task scheduling and handles them sequentially.

B.3.3 Decision Variable Method

Contrary to the previously introduced decomposition method, the decision variable method handles the problems of task decomposition, task allocation and task scheduling simultaneously. This is done by direcly modeling and considering the `AND` and `OR` operators and thus the task trees in the mathematical problem formulation used for BnP. To accomplish this, a new decision variable μ_ω is introduced for each operator $\omega \in \Omega$ with Ω being the set of all operators within the task tree of a problem instance. The binary variable μ_ω is equal to one if operator ω is fulfilled and zero otherwise. Additionally, the binary indicator variables $s^{\mathcal{N}}_{i\omega}$ and $s^{\Omega}_{\tilde{\omega}\omega}$ are needed. These are equal to one if a simple task i, or an operator $\tilde{\omega} \in \Omega$, are direct subnodes of operator ω.

The constraint

$$n_\omega \mu_\omega \leq \sum_{m \in \mathcal{K}} \sum_{r \in \mathcal{R}^m} \sum_{i \in \mathcal{N}} s^{\mathcal{N}}_{i\omega} b^m_{ir} \lambda^m_r + \sum_{\tilde{\omega} \in \Omega} s^{\Omega}_{\tilde{\omega}\omega} \mu_{\tilde{\omega}} \qquad \forall\ \omega \in \Omega \qquad \text{(C7')}$$

limits the μ_ω and must be added to the problem formulation given in Section B.2 provided that the value of n_ω is determined by

$$\texttt{OR:} \quad n_\omega := 1 \qquad \text{(B.2)}$$

$$\texttt{AND:} \quad n_\omega := \sum_{i \in \mathcal{N}} s^{\mathcal{N}}_{i\omega} + \sum_{\tilde{\omega} \in \Omega} s^{\Omega}_{\tilde{\omega}\omega} \qquad \text{(B.3)}$$

for `OR` and `AND` operators, respecitvely. The right part of (C7′) sums up how many direct subnodes of the operator ω are fulfilled. The n_ω indicates how many subnodes of operator ω must at least be fulfilled to satisfy the operator ω itself. Thus, for the fulfillment of `OR` operators, at least one subnode must be fulfilled, while for the fulfillment of `AND` operators all direct subnodes must be fulfilled.

Moreover, constraint (C2) of the original problem formulation given in Section B.2 must no longer be fulfilled if complex tasks are contained in the problem instance. It ensures all simple tasks to be executed, which is not necessary anymore as soon as `OR` operators are contained in the problem instance. For problem instances with complex dependencies, all direct subtasks and suboperators of the overall problem must be fulfilled. Let these direct subtasks and operators be contained in the sets $\mathcal{E}$ and $\mathcal{L}$, respectively. By replacing constraint (C2) with the constraints

$$\sum_{m \in \mathcal{K}} \sum_{r \in \mathcal{R}^m} b^m_{ir} \lambda^m_r = 1 \qquad \forall\ i \in \mathcal{E} \qquad \text{(C2a}'_1\text{)}$$

$$\sum_{m \in \mathcal{K}} \sum_{r \in \mathcal{R}^m} b^m_{ir} \lambda^m_r \leq 1 \qquad \forall\ i \in \mathcal{N} \setminus \mathcal{E} \qquad \text{(C2a}'_2\text{)}$$

$$\mu_\omega = 1 \qquad \forall\ \omega \in \mathcal{L} \qquad \text{(C2b}'_1\text{)}$$

$$\mu_\omega \leq 1 \qquad \forall\ \omega \in \Omega \setminus \mathcal{L}, \qquad \text{(C2b}'_2\text{)}$$

the wanted behavior is ensured. Equation (C2a$'_1$) ensures all direct subtasks of the overall problem to be fulfilled, while (C2a$'_2$) makes sure that all other elemental tasks are executed at most once. The same is ensured for the fulfillment of the operators by (C2b$'_1$) and (C2b$'_2$).

The changes in the LP require the introduction of additional branching rules. The branching rules applied for the MRTA problem instances with cross-schedule dependencies and their priorization are preserved, i. e. branching on task pairs occurring on the same route, branching on pairwise task's order for tasks allocated to the same agent's route and branching on a task being allocated to a specific agent (see Section B.2 and for more details refer to [Kor11, Chapter 5]). If none of these branching rules are applicable, the additional branching rules are applied in the following order:

Branching on the execution of an operator:
The binary decision variable μ_ω determines whether an operator ω is fulfilled or not. If the value of any μ_ω obtained from solving the problem relaxation is non-binary, one branch forces the respective operator to be fulfilled, while the other forces it to not be fulfilled, i. e.

left branch: $\mu_\omega = 1$
right branch: $\mu_\omega = 0$

In case several variables μ_ω are non-binary, the one having the value closest to 0.5 is chosen for branching, as customary.

Branching on the execution of a task:
For each simple task $i \in \mathcal{N}$, $\sum_{m \in \mathcal{K}} \sum_{r \in \mathcal{R}^m} b_{ir}^m \lambda_r^m$ determines whether a task is fully executed in a solution to the problem relaxation or not. If the value is non-binary, one branch forces the respective task to be executed, while the other forces it to not be executed, i. e.

left branch: $\sum_{m \in \mathcal{K}} \sum_{r \in \mathcal{R}^m} b_{ir}^m \lambda_r^m = 1$
right branch: $\sum_{m \in \mathcal{K}} \sum_{r \in \mathcal{R}^m} b_{ir}^m \lambda_r^m = 0$

In case several tasks are only partially executed, the task $i \in \mathcal{N}$ is considered for branching, for which $\sum_{m \in \mathcal{K}} \sum_{r \in \mathcal{R}^m} b_{ir}^m \lambda_r^m$ is closest to the value of 0.5.

One restriction to the decision variable approach is that simple task are allowed to only be used once in the task tree of the overall problem. If this restriction is neglected and a task is used multiple times in the tree, the LP sums several times over this task which can results in impermissible solutions. However, every MRTA problem with complex tasks can be formulated such that each simple task is contained only once in the overall task tree. An example of an infeasible task tree and a corresponding feasible representation of the same problem instance is depicted in Figure B.4.

B.3.4 Cluster Method

Another approach to simultaneously handle of task decomposition, task allocation, and task scheduling is the cluster method. It is based on the idea of partitioning the set of simple tasks $\mathcal{N}$ into pairwise disjoint sets of task clusters $\mathcal{C}$. The set containing

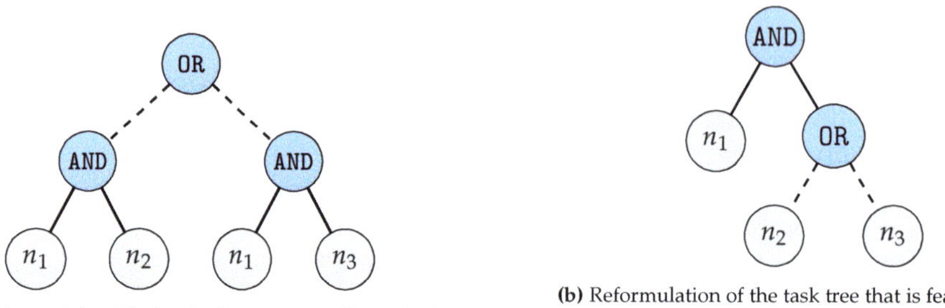

(a) Task tree infeasible for the decision variable method.

(b) Reformulation of the task tree that is feasible for the decision variable method.

Figure B.4: Example of a feasible reformulation of an infeasible task tree for the decision variable method. The figure is adapted from [Wö23].

all clusters is given by $\mathcal{C}_\Sigma$. The clusters are defined such that the junction of all clusters equal the task set, i. e.

$$\cup_{\mathcal{C}\in\mathcal{C}_\Sigma}\mathcal{C} = \mathcal{N}. \tag{B.4}$$

Each cluster $\mathcal{C}$ either contains an elemental task or all elemental tasks belonging to a complex task of the overall problem. Thus, every direct subtasks of the overall problem is stored in a separate cluster. Each elemental task is only allowed to be used in one cluster. Weight parameters are used to model possible task decompositions. Each task $n_i \in \mathcal{N}$ has a corresponding task weight g_i and each cluster $\mathcal{C} \in \mathcal{C}_\Sigma$ is allocated to a corresponding cluster weight $g_\mathcal{C}$. The weights are chosed such that for all valid minimal task decompositions

$$\sum_{m\in\mathcal{K}}\sum_{r\in\mathcal{R}^m}\sum_{i\in\mathcal{C}} g_i b_{ir}^m \lambda_r^m = g_\mathcal{C} \qquad \forall\ \mathcal{C}\in\mathcal{C}_\Sigma \tag{C2a$''$}$$

is fulfilled. Constraint (C2a$''$) ensures, that the sum of weights g_i of executed tasks equals the cluster weight $g_\mathcal{C}$ of each cluster $\mathcal{C}$.

An example of a problem with complex tasks and corresponding clusters, task and cluster weights is depicted in Figure B.5. In this example, the elemental tasks are split into three clusters $\mathcal{C}_1 = \{n_1\}$, $\mathcal{C}_2 = \{n_2\}$ and $\mathcal{C}_3 = \{n_3, n_4, n_5, n_6\}$. Clusters $\mathcal{C}_1$ and $\mathcal{C}_2$ only contain one simple task each, thus (C2a$''$) is fulfilled for these clusters with the weights being set to $g_{\mathcal{C}_1} = g_1$ and $g_{\mathcal{C}_2} = g_2$, which are all set to equal 1 in this example. The complex task represented by cluster $\mathcal{C}_3$ is fulfilled, if either task n_3, or tasks n_4 and n_6, or tasks n_5 and n_6 are executed. This is represented by setting the cluster weight to $g_{\mathcal{C}_3} = 3$ and the task weights to $g_3 = 3$, $g_4 = g_5 = 2$, and $g_6 = 1$.

As explained in the previous section, constraint (C2) that ensures the fulfillment of all simple tasks, must be neglected when considering complex tasks. In the cluster method, (C2) is replaced by constraint (C2a$''$). In the relaxed problem formulation additionally

$$\lambda_r^m \leq 1 \qquad \forall\ r\in\mathcal{R}^m, m\in\mathcal{K} \tag{C2b$''$}$$

must be considered to prevent manifold utilization of routes to fulfill (C2a$''$).

For the determination of a valid set of weight parameters, an ILP can be formulated for each cluster $\mathcal{C} \in \mathcal{C}_{\Sigma}$. The division into separate weight determination problems for each cluster is possible due to the restriction of each simple task only being contained in one cluster, i.e. the clusters being disjoint. To find the cluster and task weights for a cluster $\mathcal{C} \in \mathcal{C}_{\Sigma}$, all possible task subsets $\mathcal{B}_{\mathcal{C}} \subset \mathcal{C}$ are determined and separated into two disjoint subsets $\mathcal{B}_{\mathcal{C}}^{\mathrm{val}}$ and $\mathcal{B}_{\mathcal{C}}^{\mathrm{inv}}$. All valid minimal cluster decompositions are contained in $\mathcal{B}_{\mathcal{C}}^{\mathrm{val}}$, while $\mathcal{B}_{\mathcal{C}}^{\mathrm{inv}}$ consists of all subsets that do not form a valid minimal decomposition of the cluster $\mathcal{C}$.

The ILP seeks to find integer task and cluster weights greater zero such that the cluster weight $g_{\mathcal{C}}$ is minimal and the task weights of all valid minimal decompositions sum up to the cluster weight, i.e.

$$\min g_{\mathcal{C}} \tag{B.5}$$

$$\text{s.t.} \quad \sum_{i \in \mathcal{B}_{\mathcal{C}}} g_i = g_{\mathcal{C}} \qquad \forall\ \mathcal{B}_{\mathcal{C}} \in \mathcal{B}_{\mathcal{C}}^{\mathrm{val}}, \tag{B.6}$$

$$g_i \in \mathbb{N} \qquad \forall\ i \in \mathcal{N}, \tag{B.7}$$

$$g_{\mathcal{C}} \in \mathbb{N} \qquad \forall\ \mathcal{C} \in \mathcal{C}_{\Sigma}. \tag{B.8}$$

However, (B.7) must not be fulfilled for any invalid decomposition $\mathcal{B}_{\mathcal{C}} \in \mathcal{B}_{\mathcal{C}}^{\mathrm{inv}}$ of the cluster $\mathcal{C}$. Thus, after solving the ILP it is checked if any subset of tasks $\tilde{\mathcal{B}}_{\mathcal{C}} \in \mathcal{B}_{\mathcal{C}}^{\mathrm{inv}}$ exists, that fulfills $\sum_{i \in \tilde{\mathcal{B}}_{\mathcal{C}}} g_i = g_{\mathcal{C}}$. If this is the case, the constraint

$$\sum_{i \in \tilde{\mathcal{B}}_{\mathcal{C}}} g_i \geq g_{\mathcal{C}} + 1 \tag{B.9}$$

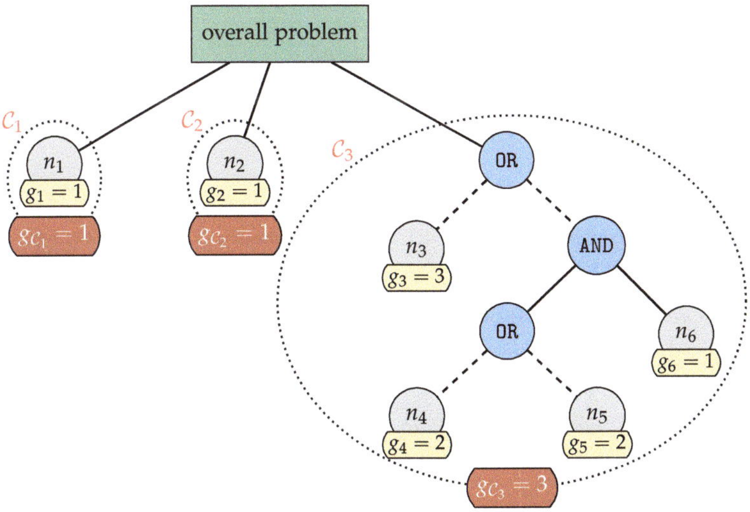

Figure B.5: Example of clusters, cluster weights and task weights in an MRTA problem instance with complex dependencies. The Figure is based on [Wö23].

is added to the ILP and it is solved anew. This process is repeated until no further $\tilde{\mathcal{B}}_{\mathcal{C}} \in \mathcal{B}_{\mathcal{C}}^{\text{inv}}$ fulfilling (B.7) can be found, which ensures the weights to be suitable.

As for the decision variable method (see Section B.3.3), also for the cluster method the changes in the problem formulation require an additional branching rule. First, the branching rules as given for the MRTA problem instances with cross-schedule dependencies and their priorization are used (see Section B.2 and for more details refer to [Kor11]). If none of these rules are applicable, the branching on the execution of a simple task as explained in the previous Section B.3.3 is applied.

B.4 Evaluation of the Proposed Methods

To compare the previously introduced decomposition method (see Section B.3.2), decision variable method (see Section B.3.3), and cluster method (see Section B.3.4), an evaluation with problem instances differing in the amount and types of complex tasks is conducted. The results of this evaluation have been published in [BWRH24].

Overall, seven different problem specifications are considered. They all contain eight simple tasks and three homogeneous agents, but differ in the amount and the type of complex tasks the simple tasks are combined to. The amount and type of complex tasks considered in the different problem specifications are depicted in Figure B.6. For each problem specification, in total ten problem instances are considered. For each problem instance, the initial positions of the agents and the positions of the simple tasks are chosen randomly over an uniform distribution of an area of the size 10×15. The basic durations of the tasks are randomly sampled from a uniform distribution over the interval $[0, 30]$. For comparability of the results, the tasks' positions and durations are equal for all problem specifications.

The results of the calculation times required by the different solution approaches are given in Table B.3.

Table B.3: Results on the calculation time in seconds (mean and standard deviation) required by the proposed BnP approaches for solving MRTA problem instances with different complex tasks.

Problem specification	Decomposition method		Decision variable method		Cluster method	
	mean	SD	mean	SD	mean	SD
A	31.77	10.57	33.03	11.22	33.47	11.55
B	8.24	1.08	25.72	4.60	16.80	4.12
C	3.21	0.25	19.72	11.74	14.89	10.13
D	2.01	0.20	15.10	9.38	11.06	7.37
E	1.19	0.05	20.54	25.93	23.04	34.54
F	1.60	0.15	19.71	8.65	333.86	709.62
G	0.30	0.01	20.77	15.38	1987.52	2330.27

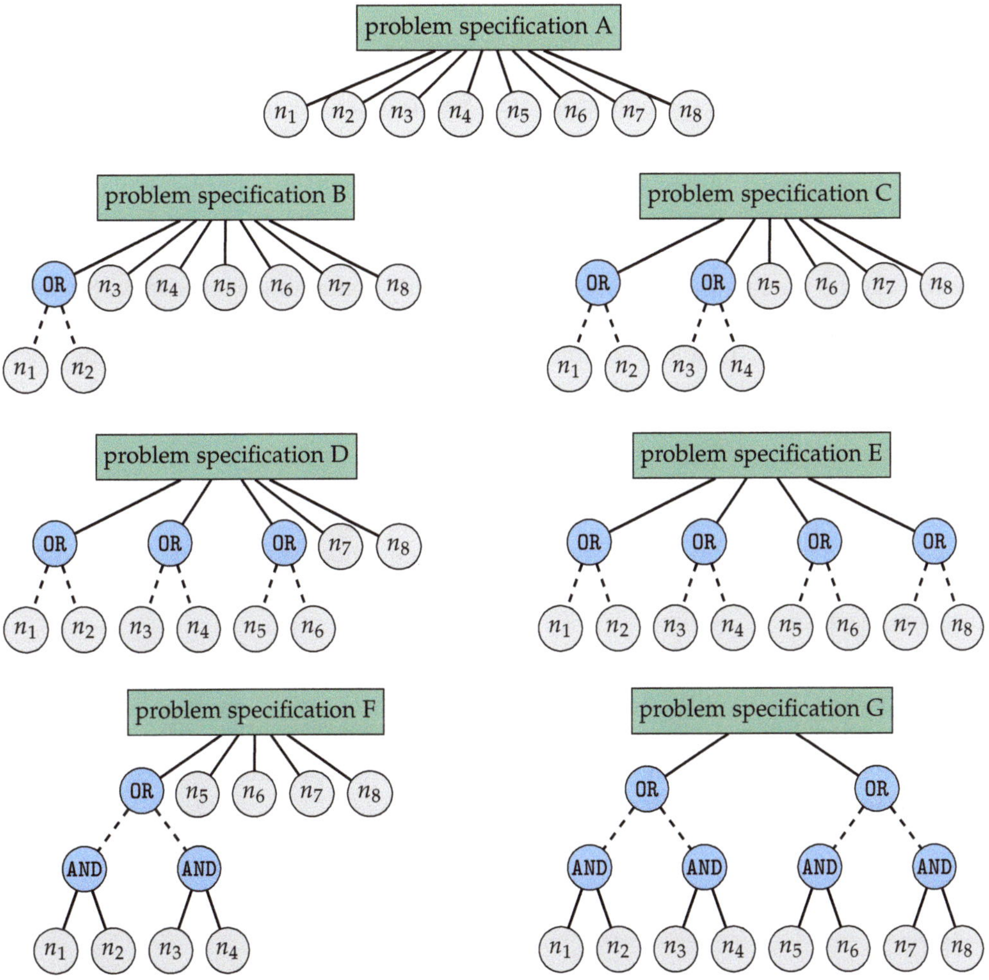

Figure B.6: Overview of the task trees of the different problem specifications considered for the evaluation of the BnP decomposition, decision variable and cluster method. The figure is based on [Wö23].

All proposed BnP methods are capable to exactly solve the examined MRTA problem instances. The calculation time required by the three methods to solve the problem instances of specification A without any complex tasks is very similar.

Independent of the amount and type of complex tasks contained in the problem instance, the decomposition method requires the least calculation time. The results indicate that the fewer tasks are simple, the less calculation time is needed by the decomposition approach to solve the respective problem instances. For this method, the influence of exponentially less calculation time needed for time-extended MRTA problem instances with fewer simple tasks (see Section 5.2) predominates the additional calculation time needed for the determination of possible minimal decompositions.

For the decision variable method and the cluster method, the mean calculation time required to solve problem instances with complex tasks that only contain two simple tasks linked by an `OR` operator (problem specifications B-E), also decreases compared to the calculation time required to solve problems only containing simple tasks (problem specification A). However, the consideration of complex tasks linking four simple tasks with more operators (problem specifications F and G) significantly increases the calculation time required by the cluster method to on average up to more than 33 min for problem specification G. In these cases, more branches are needed to solve the problem instances using the cluster approach, which results in higher calculation times. In contrast to this, the average calculation time required by the decision variable method to solve these problem instances remains approximately constant around 20 s. Thus, between these two methods, the decision variable method can be clearly favored over the cluster approach if complex tasks linking many simple tasks are contained in the problem instance.

Overall, all methods are capable of solving the considered time-extended MRTA problem instances with complex dependencies. The decomposition method clearly yields the best results and handles complex tasks the fastest. However, further optimization of the decision variable method and the cluster method probably is possible, e. g. by optimizing the order of the branching rules.

C Additional Evaluation Results

In this chapter, results supplementary to the evaluation presented in Chapter 5 are given.

C.1 Evaluation Scenario 1 – Different Sizes of Problem Instances

In this section, the additional results of evaluation scenario 1 are given. They include the results on the approximation ratio and the Levenshtein distance for the modifications of task insertion and task deletion. For both, the mean values and the standard deviations over the 50 modified problem instances of one specification are given in Tables C.1 and C.3, respectively.

Table C.1: Evaluation scenario 1: Results on the approximation ratio α (mean and standard deviation) for the modifications of task insertion and task deletion.

Variation			Reoptimization heuristic mean	Reoptimization heuristic SD	GA-based reoptimization mean	GA-based reoptimization SD	Conventional GA mean	Conventional GA SD
Task insertion	Number of tasks	2	1.0089	0.0351	1.0	0.0	1.0	0.0
		3	1.0108	0.0265	1.0	0.0	1.0	0.0
		4	1.0092	0.0196	1.0032	0.0084	1.033	0.0621
		5	1.0096	0.0181	1.0089	0.0166	1.1114	0.09
		6	1.0082	0.0173	1.0078	0.0159	1.1798	0.0969
		7	1.0078	0.0167	1.0078	0.0167	1.247	0.0786
		8	1.0078	0.0151	1.0078	0.0151	1.3117	0.0775
		9	1.0071	0.012	1.0071	0.012	1.3794	0.1054
		10	1.0071	0.0121	1.0071	0.0121	1.4342	0.0994
	Number of agents	1	1.0044	0.0132	1.0044	0.0132	1.1451	0.0564
		2	1.0053	0.0153	1.0053	0.0153	1.1848	0.062
		3	1.0043	0.0101	1.0043	0.0101	1.2537	0.0686
		4	1.0067	0.016	1.0067	0.016	1.3411	0.0833
		5	1.0065	0.0151	1.0065	0.0151	1.4004	0.1195
		6	1.0043	0.0103	1.0043	0.0103	1.4884	0.1468
		7	1.0032	0.0092	1.0032	0.0092	1.5418	0.1542
		8	1.0016	0.0045	1.0016	0.0045	1.5999	0.1659
		9	1.0009	0.0036	1.0009	0.0036	1.6512	0.1809
		10	1.0011	0.0039	1.0011	0.0039	1.6947	0.1936
Task deletion	Number of tasks	1	1.0134	0.0538	1.0	0.0	1.0	0.0
		2	1.0021	0.0091	1.0	0.0	1.0	0.0
		3	1.0049	0.0165	1.0	0.0	1.0	0.0
		4	1.0051	0.0161	1.0008	0.0038	1.0242	0.0371
		5	1.0047	0.0123	1.0026	0.008	1.0825	0.0738
		6	1.0042	0.012	1.0042	0.012	1.1584	0.0695
		7	1.0047	0.0133	1.0047	0.0133	1.2469	0.0904
		8	1.0032	0.0094	1.0032	0.0094	1.3126	0.0996
		9	1.0038	0.0102	1.0038	0.0102	1.373	0.0874
	Number of agents	1	1.0023	0.0061	1.0023	0.0061	1.1281	0.0462
		2	1.002	0.0076	1.002	0.0076	1.1763	0.0576
		3	1.0017	0.0047	1.0017	0.0047	1.2584	0.0839
		4	1.0029	0.006	1.0029	0.006	1.3406	0.098
		5	1.0038	0.0077	1.0038	0.0077	1.4081	0.1349
		6	1.0048	0.01	1.0048	0.01	1.4605	0.1395
		7	1.0084	0.0188	1.0084	0.0188	1.5196	0.1388
		8	1.0082	0.0196	1.0082	0.0196	1.5605	0.1425
		9	1.005	0.0111	1.005	0.0111	1.626	0.1704
		10	1.0037	0.0095	1.0037	0.0095	1.6663	0.1595

Table C.2: Evaluation scenario 1: Results on the calculation time in seconds (mean and standard deviation) of the for the modifications of task insertion and task deletion.

Variation			Reoptimization heuristic		GA-based reoptimization		Conventional GA		Branch-and-price	
			mean	SD	mean	SD	mean	SD	mean	SD
Task insertion	Number of tasks	2	0.0013	0.0004	10.9004	0.05	10.9161	0.1194	0.0069	0.0007
		3	0.0014	0.0043	12.3838	0.0488	12.3933	0.0587	0.0156	0.0051
		4	0.0016	0.0046	13.8013	0.0527	13.8015	0.0587	0.0327	0.0074
		5	0.0019	0.0048	15.2781	0.1023	15.2705	0.0847	0.1031	0.0249
		6	0.0025	0.0054	16.7087	0.0757	16.7185	0.0858	0.3469	0.0616
		7	0.0022	0.0053	18.0411	0.0995	18.0373	0.0949	1.4312	0.3633
		8	0.0022	0.0052	19.4094	0.0936	19.4468	0.0937	8.2996	2.593
		9	0.0023	0.0054	20.8564	0.1147	20.8121	0.1048	55.9639	21.2503
		10	0.0025	0.0055	22.2244	0.0863	22.2249	0.1048	372.5452	158.5708
	Number of agents	1	0.0019	0.0022	15.921	0.0719	15.9246	0.1182	4.3638	1.803
		2	0.0025	0.0034	17.6509	0.0809	17.661	0.0839	7.8432	3.2909
		3	0.0025	0.0057	19.3867	0.0779	19.3918	0.0823	10.1442	.9422
		4	0.0028	0.0059	21.0468	0.1196	21.0344	0.0957	12.8124	3.9262
		5	0.0025	0.0055	22.8387	0.2505	22.7884	0.2103	14.8944	4.6895
		6	0.0028	0.0057	24.5158	0.1037	24.4983	0.1292	17.6627	5.53
		7	0.0032	0.0061	26.2178	0.151	26.2237	0.1251	19.0405	6.7351
		8	0.0034	0.0051	28.1031	0.1632	28.0873	0.1753	21.1797	7.2234
		9	0.0033	0.004	29.7958	0.1522	29.8024	0.1665	21.795	7.068
		10	0.0035	0.0064	31.6498	0.2653	31.6617	0.2149	24.1618	8.2775
Task deletion	Number of tasks	1	0.0009	0.0037	9.5308	0.0423	9.5312	0.0373	0.0025	0.0054
		2	0.0013	0.004	11.0164	0.0537	11.0232	0.0518	0.0065	0.0065
		3	0.0013	0.0042	12.4531	0.05	12.4398	0.0491	0.0147	0.0066
		4	0.0016	0.0047	13.8258	0.066	13.8103	0.0487	0.0341	0.008
		5	0.0016	0.0047	15.3255	0.1074	15.3068	0.0764	0.0954	0.0217
		6	0.0016	0.0046	16.6686	0.096	16.6834	0.0757	0.3292	0.0761
		7	0.0019	0.0048	18.0936	0.0956	18.088	0.081	1.4009	0.3368
		8	0.0022	0.005	19.4686	0.09	19.4727	0.0959	8.0993	1.9977
		9	0.0025	0.0057	20.8409	0.14	20.8148	0.1726	55.4437	18.5293
	Number of agents	1	0.0019	0.0025	16.1969	0.0761	16.1922	0.0668	4.4595	1.3539
		2	0.0022	0.0052	18.0362	0.1354	18.0242	0.1025	7.8452	2.4496
		3	0.0022	0.0054	19.7478	0.0845	19.7827	0.1194	10.9295	3.3378
		4	0.0025	0.0056	21.3566	0.1015	21.3315	0.0883	14.6456	4.6009
		5	0.0025	0.0056	23.0555	0.1043	23.0463	0.1046	14.9673	4.0564
		6	0.0028	0.0031	24.9809	0.1288	24.9546	0.1051	17.9851	5.5919
		7	0.0029	0.0032	26.8527	0.108	26.8028	0.166	20.6745	6.8098
		8	0.003	0.0028	28.6156	0.13	28.6224	0.1683	21.7446	7.8269
		9	0.0033	0.0036	30.3814	0.1533	30.3969	0.1446	24.179	8.1451
		10	0.0031	0.006	32.1636	0.1524	32.1096	0.1326	25.949	8.2305

Table C.3: Evaluation scenario 1: Results on the Levenshtein distance (mean and standard deviation) for the modifications of task insertion and task deletion.

Variation			Reoptimization heuristic		GA-based reoptimization		Conventional GA		Branch-and-price	
			mean	SD	mean	SD	mean	SD	mean	SD
Task insertion	Number of tasks	2	1.0	0.0	1.24	0.6499	1.24	0.6499	1.24	0.6499
		3	1.0	0.0	1.52	1.0438	1.82	1.0713	1.74	1.0356
		4	1.0	0.0	1.78	1.6887	3.12	1.9559	2.32	1.7022
		5	1.0	0.0	1.12	0.621	5.14	2.2804	2.96	2.5997
		6	1.0	0.0	1.14	0.98	6.16	2.361	3.22	2.8516
		7	1.0	0.0	1.0	0.0	7.32	2.5569	4.26	3.8513
		8	1.0	0.0	1.0	0.0	9.22	2.5868	5.0	5.02
		9	1.0	0.0	1.0	0.0	10.48	2.8301	5.48	5.714
		10	1.0	0.0	1.0	0.0	11.82	2.7254	6.34	5.6413
	Number of agents	1	1.0	0.0	1.0	0.0	5.26	1.5338	3.5	2.7659
		2	1.0	0.0	1.0	0.0	8.64	3.242	4.78	4.5046
		3	1.0	0.0	1.0	0.0	8.98	2.4289	5.02	4.3474
		4	1.0	0.0	1.0	0.0	10.04	2.441	4.8	4.6476
		5	1.0	0.0	1.0	0.0	9.58	2.384	3.48	4.1484
		6	1.0	0.0	1.0	0.0	9.86	2.2892	3.2	3.1048
		7	1.0	0.0	1.0	0.0	10.54	2.1091	2.52	2.5631
		8	1.0	0.0	1.0	0.0	10.38	1.8643	2.36	2.0274
		9	1.0	0.0	1.0	0.0	10.76	2.1407	2.04	1.7316
		10	1.0	0.0	1.0	0.0	10.52	2.0808	2.14	1.8656
Task deletion	Number of tasks	1	1.0	0.0	1.12	0.475	1.12	0.475	1.12	0.475
		2	1.0	0.0	1.16	0.5044	1.58	0.8022	1.3	0.6403
		3	1.0	0.0	1.9	1.7464	2.44	1.6752	2.06	1.7252
		4	1.0	0.0	1.68	1.5677	3.98	2.1771	2.3	1.9723
		5	1.0	0.0	1.44	1.6144	6.1	2.3937	2.72	2.8847
		6	1.0	0.0	1.24	1.68	7.46	2.9679	3.44	3.5505
		7	1.0	0.0	1.0	0.0	9.48	2.9205	3.2	3.7736
		8	1.0	0.0	1.0	0.0	11.52	3.1064	4.4	4.4362
		9	1.0	0.0	1.0	0.0	13.24	2.8111	4.8	5.4699
	Number of agents	1	1.0	0.0	1.0	0.0	6.66	1.38	2.74	2.8622
		2	1.0	0.0	1.0	0.0	9.6	2.569	3.3	3.7216
		3	1.0	0.0	1.0	0.0	10.5	2.816	4.52	4.4955
		4	1.0	0.0	1.0	0.0	10.9	2.6401	4.6	4.5431
		5	1.0	0.0	1.0	0.0	11.4	2.8496	4.5	4.5706
		6	1.0	0.0	1.0	0.0	11.46	2.5156	4.38	4.5028
		7	1.0	0.0	1.0	0.0	12.24	2.4622	4.28	4.4228
		8	1.0	0.0	1.0	0.0	13.04	2.163	3.48	4.3965
		9	1.0	0.0	1.0	0.0	12.32	2.2578	3.32	4.1638
		10	1.0	0.0	1.0	0.0	12.7	2.2561	3.08	3.3873

C.2 Evaluation Scenario 2 – Different Heterogeneity Levels

The additional results of evaluation scenario 2 are given in this section. First, the results on the approximation rations for the modifications of task insertion, task deletion, task position variation, task duration variation, agent capability variation and agent velocity variation are given in Tables C.4 to C.7. This is followed by the results on the calculation times in Tables C.8 to C.11 and the Levenshtein distances in Tables C.12 to C.15.

Table C.4: Evaluation scenario 2: Results on the approximation ratio α (mean and standard deviation) for the modifications of task insertion and task deletion.

Mod.	Level c	Level v	Level d	Reoptimization heuristic mean	Reoptimization heuristic SD	GA-based reoptimization mean	GA-based reoptimization SD	Conventional GA mean	Conventional GA SD
Task insertion	0	0	0	1.0	0.0	1.0	0.0	1.0	0.0
	1	0	0	1.0058	0.0122	1.0058	0.0122	1.3223	0.0884
	2	0	0	1.0098	0.0306	1.0098	0.0306	1.2784	0.1028
	3	0	0	1.0049	0.0214	1.0019	0.0058	1.0882	0.0812
	0	1	0	1.0	0.0	1.0	0.0	1.0	0.0
	0	2	0	1.0012	0.0036	1.0012	0.0036	1.7153	0.2071
	2	1	0	1.0081	0.0221	1.0081	0.0221	1.28	0.1031
	2	2	0	1.0111	0.0346	1.0102	0.0312	1.3551	0.1509
	0	0	1	1.0	0.0	1.0	0.0	1.0	0.0
	0	0	2	1.0051	0.0163	1.0051	0.0163	1.2924	0.0818
	2	0	1	1.0094	0.0279	1.0094	0.0279	1.3155	0.0943
	2	0	2	1.0084	0.0242	1.0084	0.0242	1.2757	0.0884
	0	2	1	1.0012	0.0035	1.0012	0.0035	1.7538	0.164
	0	2	2	1.0009	0.0027	1.0009	0.0027	1.5848	0.17
	2	2	1	1.0097	0.0294	1.0097	0.0294	1.3511	0.1499
	2	2	2	1.0083	0.0263	1.0083	0.0263	1.3686	0.1403
Task deletion	0	0	0	1.0083	0.0147	1.0083	0.0147	1.3576	0.1097
	1	0	0	1.0055	0.0111	1.0055	0.0111	1.3268	0.1117
	2	0	0	1.0054	0.0151	1.0054	0.0151	1.2937	0.0902
	3	0	0	1.0056	0.017	1.0051	0.016	1.0806	0.0927
	0	1	0	1.0046	0.0147	1.0046	0.0147	1.4407	0.1062
	0	2	0	1.0055	0.0152	1.0055	0.0152	1.7356	0.1885
	2	1	0	1.0041	0.0106	1.0041	0.0106	1.2939	0.1023
	2	2	0	1.0052	0.0166	1.0052	0.0166	1.3495	0.1376
	0	0	1	1.0083	0.0146	1.0083	0.0146	1.3649	0.1073
	0	0	2	1.0069	0.0118	1.0069	0.0118	1.3108	0.096
	2	0	1	1.0049	0.0125	1.0049	0.0125	1.3094	0.1049
	2	0	2	1.0027	0.0077	1.0027	0.0077	1.3063	0.0897
	0	2	1	1.0055	0.0149	1.0055	0.0149	1.7532	0.1819
	0	2	2	1.0044	0.0119	1.0044	0.0119	1.6319	0.1854
	2	2	1	1.0046	0.0132	1.0046	0.0132	1.3742	0.1394
	2	2	2	1.0075	0.0183	1.0075	0.0183	1.3866	0.1364

Table C.5: Evaluation scenario 2: Results on the approximation ratio α (mean and standard deviation) for the modification of task position variation.

Mod.	Level c	Level v	Level d	DIH heuristic mean	DIH heuristic SD	INI heuristic mean	INI heuristic SD	GA-based reopt. with DIH mean	GA-based reopt. with DIH SD	GA-based reopt. with INI mean	GA-based reopt. with INI SD	Conventional GA mean	Conventional GA SD
Task position variation	0	0	0	1.0067	0.0146	1.0571	0.0697	1.0067	0.0146	1.0571	0.0697	1.3261	0.1027
	1	0	0	1.0096	0.0188	1.0632	0.0751	1.0096	0.0188	1.0632	0.0751	1.3319	0.0915
	2	0	0	1.0105	0.0226	1.0471	0.0681	1.0105	0.0226	1.0462	0.0662	1.2972	0.0925
	3	0	0	1.0058	0.0172	1.0229	0.0526	1.0057	0.0164	1.0142	0.027	1.0733	0.0762
	0	1	0	1.0062	0.0119	1.0513	0.0555	1.0062	0.0119	1.0513	0.0555	1.4122	0.1113
	0	2	0	1.005	0.0121	1.0486	0.0554	1.005	0.0121	1.0486	0.0554	1.7023	0.1625
	2	1	0	1.0064	0.0151	1.0473	0.0699	1.0064	0.0151	1.0459	0.0655	1.2912	0.0882
	2	2	0	1.004	0.0085	1.0359	0.0585	1.004	0.0085	1.0359	0.0585	1.3389	0.1198
	0	0	1	1.0067	0.0144	1.0571	0.0698	1.0067	0.0144	1.0571	0.0698	1.3401	0.1104
	0	0	2	1.0055	0.0117	1.0458	0.0568	1.0055	0.0117	1.0458	0.0568	1.2746	0.1067
	2	0	1	1.0102	0.023	1.0475	0.0707	1.0102	0.023	1.047	0.0698	1.305	0.1125
	2	0	2	1.0087	0.0221	1.0389	0.0674	1.0087	0.0221	1.0386	0.0669	1.2985	0.0869
	0	2	1	1.005	0.0121	1.0486	0.0548	1.005	0.0121	1.0486	0.0548	1.7354	0.2037
	0	2	2	1.0042	0.0101	1.0397	0.044	1.0042	0.0101	1.0389	0.0429	1.5939	0.1474
	2	2	1	1.0047	0.0106	1.0371	0.0582	1.0047	0.0106	1.0365	0.0582	1.3594	0.1556
	2	2	2	1.0032	0.0086	1.0308	0.0503	1.0032	0.0086	1.0308	0.0503	1.3741	0.1656

Table C.6: Evaluation scenario 2: Results on the approximation ratio α (mean and standard deviation) for the modification of task duration variation.

Mod.	Level c	Level v	Level d	DIH heuristic mean	DIH heuristic SD	INI heuristic mean	INI heuristic SD	GA-based reopt. with DIH mean	GA-based reopt. with DIH SD	GA-based reopt. with INI mean	GA-based reopt. with INI SD	Conventional GA mean	Conventional GA SD
Task duration variation	0	0	0	1.0	0.0	1.0	0.0	1.0	0.0	1.0	0.0	1.3387	0.1007
	1	0	0	1.0012	0.0059	1.0012	0.0059	1.0012	0.0059	1.0012	0.0059	1.3283	0.085
	2	0	0	1.0007	0.0047	1.0007	0.005	1.0007	0.0047	1.0007	0.005	1.2765	0.1055
	3	0	0	1.0006	0.0039	1.0013	0.0062	1.0006	0.0039	1.001	0.0046	1.0814	0.0758
	0	1	0	1.0	0.0	1.0	0.0	1.0	0.0	1.0	0.0	1.4374	0.1175
	0	2	0	1.0	0.0	1.0	0.0	1.0	0.0	1.0	0.0	1.7085	0.1812
	2	1	0	1.0	0.0	1.0001	0.0007	1.0	0.0	1.0001	0.0007	1.2777	0.0974
	2	2	0	1.0001	0.0006	1.0003	0.0018	1.0001	0.0006	1.0003	0.0018	1.3504	0.1288
	0	0	1	1.0	0.0	1.0	0.0	1.0	0.0	1.0	0.0	1.3503	0.0926
	0	0	2	1.0	0.0	1.0	0.0	1.0	0.0	1.0	0.0	1.2908	0.1007
	2	0	1	1.0005	0.0033	1.0005	0.0036	1.0005	0.0033	1.0005	0.0036	1.2992	0.0849
	2	0	2	1.0003	0.0019	1.0003	0.0022	1.0003	0.0019	1.0003	0.0022	1.2789	0.0992
	0	2	1	1.0	0.0	1.0	0.0	1.0	0.0	1.0	0.0	1.7654	0.195
	0	2	2	1.0	0.0	1.0	0.0	1.0	0.0	1.0	0.0	1.5986	0.1677
	2	2	1	1.0003	0.0018	1.0005	0.0025	1.0003	0.0018	1.0005	0.0025	1.3708	0.1361
	2	2	2	1.0	0.0	1.0005	0.0026	1.0	0.0	1.0005	0.0026	1.3433	0.1442

Table C.7: Evaluation scenario 2: Results on the approximation ratio α (mean and standard deviation) for the modifications of agent capability variation and agent velocity variation.

Mod.	Level			INI heuristic		GA-based reoptimization		Conventional GA	
	c	v	d	mean	SD	mean	SD	mean	SD
Agent capability variation	0	0	0	1.0	0.0	1.0	0.0	1.3232	0.1145
	1	0	0	1.005	0.016	1.005	0.016	1.3296	0.0945
	2	0	0	1.0051	0.0181	1.0051	0.0181	1.3105	0.1014
	3	0	0	1.0435	0.1015	1.0361	0.0768	1.1322	0.1348
	0	1	0	1.0023	0.0082	1.0023	0.0082	1.4151	0.1022
	0	2	0	1.0003	0.0019	1.0003	0.0019	1.7271	0.1761
	2	1	0	1.0104	0.0348	1.0104	0.0348	1.3142	0.1339
	2	2	0	1.0265	0.0703	1.0262	0.0694	1.3681	0.163
	0	0	1	1.0	0.0	1.0	0.0	1.3491	0.1161
	0	0	2	1.0	0.0	1.0	0.0	1.2938	0.0996
	2	0	1	1.0016	0.0083	1.0016	0.0083	1.3144	0.1067
	2	0	2	1.0015	0.0074	1.0015	0.0074	1.3216	0.0875
	0	2	1	1.0003	0.0012	1.0003	0.0012	1.7674	0.2062
	0	2	2	1.0019	0.0093	1.0019	0.0093	1.6341	0.1571
	2	2	1	1.0377	0.0904	1.0377	0.0904	1.4171	0.1568
	2	2	2	1.0081	0.0297	1.0081	0.0297	1.4133	0.173
Agent velocity variation	0	0	0	1.1094	0.1536	1.1048	0.1409	1.4811	0.1543
	1	0	0	1.1318	0.1658	1.1289	0.161	1.4495	0.125
	2	0	0	1.04	0.0952	1.04	0.0952	1.3318	0.1642
	3	0	0	1.0033	0.0159	1.0033	0.0159	1.0787	0.0821
	0	1	0	1.1219	0.1634	1.1183	0.1567	1.5095	0.182
	0	2	0	1.0302	0.067	1.0302	0.067	1.8024	0.3261
	2	1	0	1.0331	0.0892	1.0331	0.0892	1.352	0.1818
	2	2	0	1.0254	0.0709	1.0248	0.0687	1.4124	0.2138
	0	0	1	1.1109	0.1556	1.1042	0.1394	1.4879	0.1511
	0	0	2	1.0913	0.1287	1.0853	0.1159	1.3903	0.1454
	2	0	1	1.029	0.0741	1.029	0.074	1.3386	0.1614
	2	0	2	1.0193	0.0596	1.0185	0.0585	1.3336	0.1432
	0	2	1	1.0299	0.0658	1.0299	0.0658	1.8244	0.2888
	0	2	2	1.0236	0.0509	1.0236	0.0509	1.6257	0.2061
	2	2	1	1.0252	0.0802	1.0252	0.0802	1.3956	0.1726
	2	2	2	1.0061	0.0282	1.0061	0.0282	1.3927	0.1657

Table C.8: Evaluation scenario 2: Results on the calculation time in seconds (mean and standard deviation) for the modifications of task insertion and task deletion.

Mod.	Level c	Level v	Level d	Reoptimization heuristic mean	Reoptimization heuristic SD	GA-based reoptimization mean	GA-based reoptimization SD	Conventional GA mean	Conventional GA SD	Branch-and-price mean	Branch-and-price SD
Task insertion	0	0	0	0.0018	0.0048	19.5509	0.2149	19.547	0.1211	7.4716	2.7478
	1	0	0	0.0027	0.0056	19.6815	0.1033	19.6993	0.0993	8.2547	2.7195
	2	0	0	0.003	0.0061	19.7718	0.1248	19.7833	0.133	3.4833	1.4647
	3	0	0	0.0028	0.0059	19.8405	0.145	19.8314	0.1299	0.8641	0.9172
	0	1	0	0.0018	0.0048	19.8157	0.1367	19.8131	0.1175	7.1199	2.2529
	0	2	0	0.0023	0.0041	19.8423	0.1077	19.8121	0.1129	4.9871	1.8414
	2	1	0	0.002	0.0049	19.7807	0.1118	19.7582	0.1244	3.5387	1.6199
	2	2	0	0.0032	0.0115	19.5568	0.1377	19.5616	0.1238	2.8643	1.1284
	0	0	1	0.003	0.0058	19.6011	0.1872	19.6192	0.141	7.8474	2.6915
	0	0	2	0.002	0.005	19.6335	0.1201	19.6388	0.1302	9.3631	3.7387
	2	0	1	0.0029	0.0056	20.1354	0.1556	20.1432	0.1906	3.4105	1.1839
	2	0	2	0.0027	0.0057	20.2188	0.1475	20.2094	0.1797	4.1998	1.4018
	0	2	1	0.0022	0.005	20.1099	0.1211	20.1329	0.144	5.2483	2.1102
	0	2	2	0.0022	0.0051	20.2562	0.131	20.2468	0.1368	6.4756	2.6726
	2	2	1	0.0019	0.0048	20.228	0.1116	20.2579	0.1024	2.9706	1.309
	2	2	2	0.002	0.0051	20.2004	0.1668	20.2002	0.1531	3.9252	1.2333
Task deletion	0	0	0	0.0014	0.0038	19.5194	0.1834	19.5298	0.1221	7.7449	3.2936
	1	0	0	0.0025	0.0053	19.6969	0.1256	19.7175	0.2361	7.3883	1.9772
	2	0	0	0.0014	0.0042	19.7675	0.1089	19.7468	0.1096	3.1314	1.0324
	3	0	0	0.0022	0.0051	19.8433	0.2046	19.803	0.1048	0.5246	0.527
	0	1	0	0.0023	0.0054	19.814	0.1141	19.8109	0.17	7.196	2.1659
	0	2	0	0.0027	0.0041	19.8001	0.1053	19.8248	0.1223	5.4208	1.5175
	2	1	0	0.0022	0.0054	19.7508	0.1166	19.7545	0.107	3.2989	1.1706
	2	2	0	0.002	0.0048	19.5336	0.1052	19.5508	0.123	3.0041	1.2462
	0	0	1	0.0013	0.0041	19.5911	0.1137	19.5942	0.1262	7.5072	2.3842
	0	0	2	0.0041	0.012	19.657	0.1159	19.6313	0.1451	9.1899	3.5854
	2	0	1	0.0019	0.0051	20.1006	0.1672	20.1464	0.1779	3.1474	1.1823
	2	0	2	0.0021	0.0051	20.2136	0.1419	20.2353	0.1519	3.962	1.2307
	0	2	1	0.0023	0.0051	20.1149	0.1268	20.1488	0.2381	5.3867	1.7477
	0	2	2	0.0044	0.012	20.2183	0.1027	20.2359	0.1756	7.0591	3.0239
	2	2	1	0.0021	0.0051	20.2182	0.145	20.2527	0.1273	2.9589	1.2406
	2	2	2	0.0025	0.0055	20.1906	0.153	20.1983	0.1215	3.8086	1.1598

Table C.9: Evaluation scenario 2: Results on the calculation time in seconds (mean and standard deviation) for the modification of task position variation.

Mod.	Level			DIH heuristic		INI heuristic		GA-based reopt. with DIH		GA-based reopt. with INI		Conventional GA		Branch-and-price	
	c	v	d	mean	SD	mean	SD	mean	SD	mean	SD	mean	SD	mean	SD
Task position variation	0	0	0	0.0048	0.0068	0.0034	0.0063	19.5261	0.1031	19.5163	0.1195	19.531	0.1162	7.9584	2.3851
	1	0	0	0.0036	0.0061	0.0018	0.0047	19.6974	0.0935	19.6449	0.0951	19.6855	0.0983	8.6492	2.6216
	2	0	0	0.0031	0.0062	0.0032	0.0061	19.7699	0.0894	19.7856	0.1066	19.7842	0.1093	3.1495	1.0826
	3	0	0	0.0041	0.0068	0.0024	0.0055	19.8096	0.1155	19.8415	0.1083	19.8159	0.1255	0.7105	0.8992
	0	1	0	0.0059	0.0124	0.0025	0.0057	19.7906	0.1174	19.7958	0.1146	19.7891	0.1119	7.2826	2.0259
	0	2	0	0.0038	0.0052	0.0032	0.0106	19.822	0.2034	19.8506	0.1565	19.815	0.1158	5.1822	1.6548
	2	1	0	0.004	0.0065	0.0024	0.0053	19.7448	0.1286	19.745	0.1085	19.7865	0.2288	3.376	1.1757
	2	2	0	0.0034	0.0058	0.0028	0.0055	19.5429	0.1164	19.5207	0.1278	19.5277	0.1047	3.1101	1.1323
	0	0	1	0.0049	0.007	0.0024	0.0053	19.6081	0.1043	19.6736	0.3482	19.6354	0.1632	8.1114	2.4378
	0	0	2	0.0046	0.0069	0.002	0.005	19.6473	0.1464	19.6492	0.1588	19.6298	0.0834	9.7915	3.3737
	2	0	1	0.006	0.0109	0.002	0.0051	20.0825	0.1441	20.1413	0.2172	20.1187	0.1709	3.4315	1.2444
	2	0	2	0.0047	0.007	0.002	0.0051	20.2254	0.1344	20.2521	0.1542	20.1926	0.1387	4.3473	1.4569
	0	2	1	0.0039	0.0062	0.0032	0.0058	20.0934	0.1071	20.0949	0.118	20.1092	0.105	5.4368	1.7139
	0	2	2	0.0043	0.0067	0.0011	0.0036	20.2593	0.141	20.2129	0.1133	20.2705	0.225	6.6701	2.2263
	2	2	1	0.0042	0.0067	0.0022	0.0054	20.2495	0.113	20.2373	0.1199	20.1992	0.1266	3.2069	1.5318
	2	2	2	0.0056	0.0073	0.0016	0.0047	20.2122	0.1158	20.2124	0.1457	20.1851	0.142	4.1095	1.209

Table C.10: Evaluation scenario 2: Results on the calculation time in seconds (mean and standard deviation) for the modification of task duration variation.

Mod.	Level c	Level v	Level d	DIH heuristic mean	DIH heuristic SD	INI heuristic mean	INI heuristic SD	GA-based reopt. with DIH mean	GA-based reopt. with DIH SD	GA-based reopt. with INI mean	GA-based reopt. with INI SD	Conventional GA mean	Conventional GA SD	Branch-and-price mean	Branch-and-price SD
Task duration variation	0	0	0	0.0059	0.0075	0.0014	0.0038	19.5062	0.0995	19.5087	0.1073	19.5255	0.1073	7.5067	2.2915
	1	0	0	0.0042	0.0066	0.0028	0.0056	19.6846	0.0966	19.6915	0.1115	19.7007	0.0955	8.5191	2.3673
	2	0	0	0.0037	0.0065	0.0031	0.006	19.766	0.0995	19.7799	0.1061	19.7679	0.1234	3.6616	1.3735
	3	0	0	0.0046	0.007	0.0019	0.0046	19.8222	0.1108	19.8312	0.1243	19.8108	0.1375	0.7378	0.9946
	0	1	0	0.0052	0.0071	0.0018	0.0048	19.8024	0.1367	19.7837	0.1073	19.8655	0.3107	7.143	2.2115
	0	2	0	0.0034	0.0044	0.002	0.0039	19.8367	0.1092	19.8225	0.1198	19.7991	0.1114	5.0668	1.4212
	2	1	0	0.0053	0.0073	0.0015	0.0044	19.7646	0.1088	19.8755	0.7073	19.7461	0.1106	3.5769	1.6092
	2	2	0	0.0046	0.0067	0.002	0.0048	19.5351	0.1102	19.6001	0.3875	19.5394	0.1085	3.3709	1.4947
	0	0	1	0.0041	0.0063	0.0016	0.0047	19.5829	0.1015	19.6375	0.1281	19.6094	0.1102	7.5589	2.0451
	0	0	2	0.0041	0.0067	0.0015	0.0045	19.6552	0.1105	19.6461	0.178	19.6702	0.1979	9.0524	3.0947
	2	0	1	0.0056	0.0072	0.0017	0.0046	20.1636	0.3141	20.171	0.4553	20.1627	0.4976	3.7682	1.3325
	2	0	2	0.0049	0.0069	0.0018	0.0045	20.238	0.1407	20.2536	0.1536	20.2357	0.1653	4.5276	1.5566
	0	2	1	0.0051	0.0103	0.0026	0.0054	20.1168	0.1194	20.1121	0.1065	20.1067	0.1258	5.1572	1.4974
	0	2	2	0.0053	0.0071	0.0013	0.0042	20.2456	0.1839	20.201	0.0979	20.2741	0.1122	6.5357	2.1128
	2	2	1	0.0047	0.0069	0.0024	0.0055	20.2266	0.1456	20.245	0.2035	20.2523	0.1272	3.2877	1.3111
	2	2	2	0.0058	0.0122	0.0016	0.0047	20.1842	0.1388	20.2109	0.1511	20.1679	0.1585	4.0817	1.6441

Table C.11: Evaluation scenario 2: Results on the calculation time in seconds (mean and standard deviation) for the modifications of agent capability variation and agent velocity variation.

Mod.	Level c	Level v	Level d	INI heuristic mean	INI heuristic SD	GA-based reoptimization mean	GA-based reoptimization SD	Conventional GA mean	Conventional GA SD	Branch-and-price mean	Branch-and-price SD
Agent capability variation	0	0	0	0.0009	0.0033	19.6756	0.2698	19.6742	0.1739	7.7664	2.0276
	1	0	0	0.002	0.0051	19.6903	0.1074	19.6725	0.1051	8.3662	2.5361
	2	0	0	0.0021	0.0053	19.7917	0.1076	19.7878	0.0973	4.4324	1.8518
	3	0	0	0.0025	0.0055	19.838	0.1087	19.8123	0.1319	1.2896	1.3567
	0	1	0	0.0013	0.0042	19.7859	0.12	19.7963	0.1212	7.3291	2.5487
	0	2	0	0.0029	0.005	19.8562	0.1998	19.8335	0.1147	5.1415	1.541
	2	1	0	0.001	0.0035	19.726	0.0952	19.7743	0.1457	4.1687	1.9139
	2	2	0	0.0024	0.0052	19.5368	0.1434	19.5387	0.1199	3.7333	1.6295
	0	0	1	0.0012	0.0042	19.6051	0.1272	19.6057	0.1175	7.9828	2.9819
	0	0	2	0.0026	0.0057	19.6651	0.1165	19.6306	0.1209	9.3862	3.4826
	2	0	1	0.0014	0.0044	20.1446	0.3216	20.1396	0.2283	4.2045	1.7895
	2	0	2	0.0019	0.0052	20.2593	0.156	20.2387	0.1609	5.2096	1.9235
	0	2	1	0.0017	0.0047	20.1046	0.1003	20.1228	0.1099	5.0016	1.4141
	0	2	2	0.0024	0.0053	20.2075	0.132	20.2349	0.1373	6.5013	1.9478
	2	2	1	0.0016	0.0044	20.2412	0.1199	20.251	0.112	4.1159	1.863
	2	2	2	0.0017	0.0047	20.1826	0.1293	20.2	0.1261	5.098	3.0656
Agent velocity variation	0	0	0	0.0025	0.0052	19.5416	0.1483	19.5378	0.1404	6.8959	2.2569
	1	0	0	0.0014	0.0042	19.6825	0.1157	19.7104	0.1175	7.0591	2.1465
	2	0	0	0.0014	0.0044	19.7827	0.0836	19.8109	0.1186	3.5822	1.5602
	3	0	0	0.0006	0.0028	19.8304	0.1218	19.8106	0.1139	0.7924	1.1997
	0	1	0	0.0021	0.0048	19.8014	0.1001	19.8242	0.1196	6.7766	1.9872
	0	2	0	0.0028	0.005	19.8273	0.0961	19.7952	0.1051	4.8971	1.5333
	2	1	0	0.0027	0.0058	19.7802	0.1122	19.7467	0.1151	3.435	1.7146
	2	2	0	0.002	0.0047	19.5631	0.0986	19.5554	0.1431	2.9599	1.006
	0	0	1	0.004	0.0102	19.5931	0.119	19.6048	0.1156	7.3226	2.6144
	0	0	2	0.0017	0.0047	19.7033	0.3094	19.647	0.1222	9.358	3.5496
	2	0	1	0.0014	0.0043	20.1307	0.1913	20.1528	0.194	3.4044	1.2675
	2	0	2	0.0018	0.0048	20.2257	0.1714	20.2357	0.1394	4.6972	2.681
	0	2	1	0.0021	0.0051	20.0821	0.1121	20.1101	0.1261	5.0114	1.7513
	0	2	2	0.0022	0.0051	20.2548	0.1482	20.2777	0.2308	6.2563	2.1298
	2	2	1	0.0023	0.0053	20.2513	0.1365	20.2573	0.1251	3.4895	1.8147
	2	2	2	0.0021	0.0052	20.2383	0.1401	20.1892	0.1269	4.176	1.7741

Table C.12: Evaluation scenario 2: Results on the Levenshtein distance (mean and standard deviation) for the modifications of task insertion and task deletion.

Mod.	Level c	v	d	Reoptimization heuristic mean	SD	GA-based reoptimization mean	SD	Conventional GA mean	SD	Branch-and-price mean	SD
Task insertion	0	0	0	1.0	0.0	1.0	0.0	9.18	2.8614	3.72	4.0745
	1	0	0	1.0	0.0	1.0	0.0	9.54	2.7583	5.22	4.5223
	2	0	0	1.0	0.0	1.0	0.0	8.24	2.2677	4.28	3.8473
	3	0	0	1.0	0.0	1.12	0.84	5.56	1.7454	2.48	2.1377
	0	1	0	1.0	0.0	1.0	0.0	9.66	2.3031	2.76	3.338
	0	2	0	1.0	0.0	1.0	0.0	8.24	2.294	2.76	2.6348
	2	1	0	1.0	0.0	1.0	0.0	7.72	1.9083	4.14	3.6933
	2	2	0	1.0	0.0	1.18	1.26	7.66	1.976	2.84	2.5248
	0	0	1	1.0	0.0	1.0	0.0	9.5	2.8302	3.78	4.0264
	0	0	2	1.0	0.0	1.0	0.0	8.96	3.1746	4.18	3.9633
	2	0	1	1.0	0.0	1.0	0.0	7.74	2.2344	4.0	3.5944
	2	0	2	1.0	0.0	1.0	0.0	7.96	2.0392	3.32	3.379
	0	2	1	1.0	0.0	1.0	0.0	8.38	1.875	2.46	2.4836
	0	2	2	1.0	0.0	1.0	0.0	8.9	1.9	2.64	2.575
	2	2	1	1.0	0.0	1.0	0.0	7.46	1.8353	3.14	3.0332
	2	2	2	1.0	0.0	1.0	0.0	7.38	1.9687	3.06	3.2212
Task deletion	0	0	0	1.0	0.0	1.0	0.0	10.08	2.8484	4.74	4.4758
	1	0	0	1.0	0.0	1.0	0.0	10.54	2.4997	5.16	5.2129
	2	0	0	1.0	0.0	1.0	0.0	9.92	2.2525	3.14	3.3106
	3	0	0	1.0	0.0	1.16	0.88	5.74	2.0766	3.28	2.6156
	0	1	0	1.0	0.0	1.0	0.0	11.14	2.8845	3.58	3.4934
	0	2	0	1.0	0.0	1.0	0.0	10.22	2.0812	3.4	3.1749
	2	1	0	1.0	0.0	1.0	0.0	10.32	3.0948	3.06	3.101
	2	2	0	1.0	0.0	1.0	0.0	9.08	2.2257	4.54	3.8637
	0	0	1	1.0	0.0	1.0	0.0	9.94	2.716	4.8	4.4091
	0	0	2	1.0	0.0	1.0	0.0	10.7	2.6401	4.48	4.392
	2	0	1	1.0	0.0	1.0	0.0	8.72	2.2004	4.2	3.7148
	2	0	2	1.0	0.0	1.0	0.0	9.56	2.0607	2.26	2.4067
	0	2	1	1.0	0.0	1.0	0.0	10.52	2.3853	3.3	3.1448
	0	2	2	1.0	0.0	1.0	0.0	10.56	2.4262	3.24	3.0434
	2	2	1	1.0	0.0	1.0	0.0	8.68	2.1581	3.3	3.1702
	2	2	2	1.0	0.0	1.0	0.0	9.2	1.8868	3.82	3.5704

Table C.13: Evaluation scenario 2: Results on the Levenshtein distance (mean and standard deviation) for the modification of task position variation.

Mod.	Level c	Level v	Level d	DIH heuristic mean	DIH heuristic SD	INI heuristic mean	INI heuristic SD	GA-based reopt. with DIH mean	GA-based reopt. with DIH SD	GA-based reopt. with INI mean	GA-based reopt. with INI SD	Conventional GA mean	Conventional GA SD	Branch-and-price mean	Branch-and-price SD
Task position variation	0	0	0	1.2	0.9798	0.0	0.0	1.2	0.9798	0.0	0.0	10.08	2.6821	4.3	4.6615
	1	0	0	1.12	0.9928	0.0	0.0	1.12	0.9928	0.0	0.0	9.48	2.9614	5.62	4.0541
	2	0	0	0.76	0.9708	0.0	0.0	0.76	0.9708	0.28	1.4288	9.04	2.6755	4.22	3.8538
	3	0	0	0.6	0.9165	0.0	0.0	0.72	1.1839	0.86	2.1634	5.4	1.9494	2.32	2.2221
	0	1	0	1.28	0.96	0.0	0.0	1.28	0.96	0.0	0.0	10.78	3.1196	4.68	4.5758
	0	2	0	1.32	0.9474	0.0	0.0	1.32	0.9474	0.0	0.0	9.58	1.7787	3.5	3.189
	2	1	0	0.76	0.9708	0.0	0.0	0.76	0.9708	0.2	1.4	8.7	2.508	2.58	3.0925
	2	2	0	0.68	0.9474	0.0	0.0	0.68	0.9474	0.0	0.0	8.28	2.3583	4.46	3.8585
	0	0	1	1.2	0.9798	0.0	0.0	1.2	0.9798	0.0	0.0	10.22	2.5479	4.92	4.4937
	0	0	2	1.24	0.9708	0.0	0.0	1.24	0.9708	0.0	0.0	9.12	2.9301	4.54	4.5879
	2	0	1	0.8	0.9798	0.0	0.0	0.8	0.9798	0.18	1.26	8.7	2.5865	4.06	3.8803
	2	0	2	0.8	0.9798	0.0	0.0	0.8	0.9798	0.16	1.12	8.32	1.9125	3.26	3.4975
	0	2	1	1.32	0.9474	0.0	0.0	1.32	0.9474	0.0	0.0	9.34	2.0553	3.44	3.0735
	0	2	2	1.32	0.9474	0.0	0.0	1.32	0.9474	0.08	0.56	9.04	1.5869	3.54	3.1062
	2	2	1	0.68	0.9474	0.0	0.0	0.68	0.9474	0.18	1.26	7.92	2.1985	3.78	3.684
	2	2	2	0.72	0.96	0.0	0.0	0.72	0.96	0.0	0.0	8.5	1.9209	3.06	3.3431

Table C.14: Evaluation scenario 2: Results on the Levenshtein distance (mean and standard deviation) for the modification of task duration variation.

Mod.	Level c	Level v	Level d	DIH heuristic mean	DIH heuristic SD	INI heuristic mean	INI heuristic SD	GA-based reopt. with DIH mean	GA-based reopt. with DIH SD	GA-based reopt. with INI mean	GA-based reopt. with INI SD	Conventional GA mean	Conventional GA SD	Branch-and-price mean	Branch-and-price SD
Task duration variation	0	0	0	0.04	0.28	0.0	0.0	0.04	0.28	0.0	0.0	9.64	2.7185	0.28	1.3862
	1	0	0	0.0	0.0	0.0	0.0	0.0	0.0	0.0	0.0	9.98	2.5337	1.26	3.7138
	2	0	0	0.04	0.28	0.0	0.0	0.04	0.28	0.0	0.0	9.3	2.7659	0.68	2.4855
	3	0	0	0.2	0.6	0.0	0.0	0.36	1.0346	0.3	1.3	5.42	2.108	0.56	1.7339
	0	1	0	0.0	0.0	0.0	0.0	0.0	0.0	0.0	0.0	10.32	2.7014	0.16	1.12
	0	2	0	0.0	0.0	0.0	0.0	0.0	0.0	0.0	0.0	9.48	1.8787	0.32	1.5677
	2	1	0	0.08	0.3919	0.0	0.0	0.08	0.3919	0.0	0.0	8.56	2.6993	0.16	0.88
	2	2	0	0.08	0.3919	0.0	0.0	0.08	0.3919	0.0	0.0	8.22	2.0716	0.78	2.0522
	0	0	1	0.04	0.28	0.0	0.0	0.04	0.28	0.0	0.0	10.08	3.4919	0.16	0.88
	0	0	2	0.04	0.28	0.0	0.0	0.04	0.28	0.0	0.0	9.9	2.7803	0.24	1.0307
	2	0	1	0.04	0.28	0.0	0.0	0.04	0.28	0.0	0.0	8.78	2.4681	1.56	3.2812
	2	0	2	0.04	0.28	0.0	0.0	0.04	0.28	0.0	0.0	7.94	1.8805	1.28	3.0136
	0	2	1	0.0	0.0	0.0	0.0	0.0	0.0	0.0	0.0	9.38	2.0581	0.76	2.294
	0	2	2	0.0	0.0	0.0	0.0	0.0	0.0	0.0	0.0	9.26	2.2344	0.48	1.8999
	2	2	1	0.08	0.3919	0.0	0.0	0.08	0.3919	0.0	0.0	8.5	2.184	1.2	2.735
	2	2	2	0.12	0.475	0.0	0.0	0.12	0.475	0.0	0.0	7.36	1.8736	0.92	2.5047

Table C.15: Evaluation scenario 2: Results on the Levenshtein distance (mean and standard deviation) for the modifications of agent capability variation and agent velocity variation.

Mod.	Level c	Level v	Level d	INI heuristic mean	INI heuristic SD	GA-based reoptimization mean	GA-based reoptimization SD	Conventional GA mean	Conventional GA SD	Branch-and-price mean	Branch-and-price SD
Agent capability variation	0	0	0	0.0	0.0	0.0	0.0	9.5	2.8513	2.28	3.2499
	1	0	0	0.0	0.0	0.0	0.0	10.08	3.1102	3.98	6.0216
	2	0	0	0.0	0.0	0.0	0.0	8.36	2.5904	1.36	3.7135
	3	0	0	0.0	0.0	0.6	2.4819	6.8	2.4	2.38	3.5094
	0	1	0	0.0	0.0	0.0	0.0	10.76	2.7317	2.76	5.1983
	0	2	0	0.0	0.0	0.0	0.0	9.0	1.8974	0.84	2.301
	2	1	0	0.0	0.0	0.0	0.0	9.26	3.2051	2.12	4.7229
	2	2	0	0.0	0.0	0.24	1.68	8.48	2.256	2.98	5.4863
	0	0	1	0.0	0.0	0.0	0.0	10.36	2.9649	1.36	2.784
	0	0	2	0.0	0.0	0.0	0.0	10.72	2.8428	2.16	3.1455
	2	0	1	0.0	0.0	0.0	0.0	9.22	2.7371	1.64	3.9332
	2	0	2	0.0	0.0	0.0	0.0	8.34	2.0747	1.16	2.8661
	0	2	1	0.0	0.0	0.0	0.0	9.06	1.9017	1.08	2.6895
	0	2	2	0.0	0.0	0.0	0.0	9.44	1.8128	1.92	3.4167
	2	2	1	0.0	0.0	0.0	0.0	8.52	2.36	3.08	5.6704
	2	2	2	0.0	0.0	0.0	0.0	8.74	1.9059	1.46	4.1338
Agent velocity variation	0	0	0	0.0	0.0	0.9	3.1512	10.2	2.1726	7.56	6.6578
	1	0	0	0.0	0.0	0.72	2.8708	10.44	2.7434	7.28	6.8616
	2	0	0	0.0	0.0	0.0	0.0	9.16	2.8869	3.9	5.3413
	3	0	0	0.0	0.0	0.0	0.0	5.52	1.6154	2.06	2.6938
	0	1	0	0.0	0.0	0.28	1.96	10.94	2.8873	6.96	6.8701
	0	2	0	0.0	0.0	0.0	0.0	9.16	2.2747	5.08	6.2477
	2	1	0	0.0	0.0	0.0	0.0	9.06	2.9354	3.26	4.9712
	2	2	0	0.0	0.0	0.26	1.82	8.02	2.1493	2.88	4.6632
	0	0	1	0.0	0.0	0.58	2.8572	10.12	2.84	7.76	6.5652
	0	0	2	0.0	0.0	0.28	1.96	10.04	2.6755	7.76	6.6259
	2	0	1	0.0	0.0	0.2	1.4	8.18	2.3468	4.2	5.1962
	2	0	2	0.0	0.0	0.22	1.54	8.16	1.88	3.28	5.0082
	0	2	1	0.0	0.0	0.0	0.0	9.64	1.7408	5.42	6.2067
	0	2	2	0.0	0.0	0.0	0.0	9.3	2.5788	5.1	6.265
	2	2	1	0.0	0.0	0.0	0.0	8.0	2.126	2.06	4.0714
	2	2	2	0.0	0.0	0.0	0.0	7.8	1.7436	1.72	3.6444

C.3 Evaluation Scenario 3 – Temporal Constraints

A depiction of the additional results of evaluation scenario 3 is given in this section. They contain the results for the modifications of task insertion, task deletion, task position variation, task duration variation, agent capability variation, agent velocity variation, precedence constraint insertion, synchronization constraint insertion, precedence constraint deletion and synchronization constraint deletion. The results on the approximation ratios are given in Tables C.16 to C.18 followed by the results on the calculation times in Tables C.19 to C.21 and on the Levenshtein distances in Tables C.22 to C.24.

Table C.16: Evaluation scenario 3: Results on the approximation ratio α (mean and standard deviation) for the modifications of agent capability variation and agent velocity variation.

Mod.	Constraints in $\mathcal{I}$ sync.	prec.	INI heuristic mean	SD	GA-based reopt. with INI mean	SD	Conventional GA mean	SD
Agent capability variation	0	0	1.0	0.0	1.0	0.0	1.3232	0.1145
	0	1	1.0002	0.0011	1.0002	0.0011	1.3032	0.1033
	0	2	1.0003	0.0019	1.0003	0.0019	1.3465	0.0945
	1	0	1.0013	0.0054	1.0013	0.0054	1.2184	0.0747
	2	0	1.0012	0.0041	1.0012	0.0041	1.2481	0.0971
	1	1	1.0008	0.0027	1.0006	0.0024	1.2747	0.0928
Agent velocity variation	0	0	1.1094	0.1536	1.1048	0.1409	1.4811	0.1543
	0	1	1.1134	0.1562	1.0934	0.1291	1.4011	0.1549
	0	2	1.1326	0.1698	1.1293	0.1621	1.4426	0.177
	1	0	1.1136	0.1387	1.1005	0.1149	1.2374	0.1028
	2	0	1.0884	0.0844	1.0804	0.0746	1.2361	0.1043
	1	1	1.1031	0.1478	1.0903	0.1147	1.3455	0.1374

Table C.17: Evaluation scenario 3: Results on the approximation ratio α (mean and standard deviation) for the modifications of task position variation and task duration variation.

Mod.	Constraints in $\mathcal{I}$		DIH heuristic		INI heuristic		GA-based reopt. with DIH		GA-based reopt. with INI		Conventional GA	
	sync.	prec.	mean	SD	mean	SD	mean	SD	mean	SD	mean	SD
Task position variation	0	0	1.0067	0.0146	1.0571	0.0697	1.0067	0.0146	1.0571	0.0697	1.3261	0.1027
	0	1	1.0302	0.1076	1.0566	0.0696	1.0192	0.0573	1.0526	0.067	1.3034	0.1057
	0	2	1.0227	0.0765	1.0536	0.0727	1.0212	0.0671	1.0536	0.0727	1.3162	0.1046
	1	0	1.0421	0.0729	1.0577	0.073	1.0417	0.0719	1.0575	0.0729	1.2219	0.0896
	2	0	1.0639	0.1043	1.0606	0.0604	1.0467	0.0568	1.0546	0.0513	1.2139	0.0899
	1	1	1.0393	0.0492	1.0591	0.0821	1.0388	0.0483	1.0549	0.0693	1.2707	0.0914
Task duration variation	0	0	1.0	0.0	1.0	0.0	1.0	0.0	1.0	0.0	1.3387	0.1007
	0	1	1.0	0.0	1.0	0.0	1.0	0.0	1.0	0.0	1.3025	0.1168
	0	2	1.0	0.0	1.0	0.0	1.0	0.0	1.0	0.0	1.3369	0.1284
	1	0	1.0151	0.0303	1.0001	0.0004	1.0151	0.0303	1.0001	0.0004	1.21	0.0777
	2	0	1.026	0.0426	1.0008	0.0029	1.022	0.0318	1.0008	0.0029	1.2129	0.0995
	1	1	1.0187	0.0477	1.0005	0.0022	1.0168	0.0396	1.0005	0.0022	1.2822	0.1054

Table C.18: Evaluation scenario 3: Results on the approximation ratio α (mean and standard deviation) for the modifications of precedence constraint insertion, synchronization constraint insertion, precedence constraint deletion and synchronization constraint deletion.

Mod.	Constraints in $\mathcal{I}$ sync.	prec.	Reoptimization heuristic mean	SD	GA-based reoptimization mean	SD	Conventional GA mean	SD
Precedence constr. insertion	0	0	1.0412	0.0459	1.0412	0.0459	1.3043	0.1087
	0	1	1.0313	0.0781	1.0302	0.0737	1.3338	0.1007
	0	2	1.0316	0.0495	1.0316	0.0495	1.3633	0.128
	1	0	1.0358	0.0732	1.0354	0.0717	1.2729	0.0852
	2	0	1.0425	0.0913	1.0404	0.0867	1.3118	0.1131
	1	1	1.0252	0.0464	1.0252	0.0464	1.3252	0.0857
Synchronization constr. insertion	0	0	1.1837	0.1348	1.1278	0.0786	1.2161	0.088
	0	1	1.1611	0.1434	1.1263	0.0822	1.2957	0.1148
	0	2	1.1792	0.1927	1.1443	0.1329	1.3405	0.16
	1	0	1.1	0.0848	1.0863	0.0671	1.2232	0.0904
	2	0	1.1162	0.1085	1.1039	0.093	1.2556	0.1215
	1	1	1.1021	0.1085	1.0944	0.0842	1.2872	0.1146
Precedence constr. deletion	0	0						
	0	1	1.0042	0.0199	1.0042	0.0199	1.3217	0.1066
	0	2	1.0179	0.0319	1.0179	0.0319	1.3081	0.103
	1	0						
	2	0						
	1	1	1.0196	0.0352	1.0196	0.0352	1.2204	0.0985
Synchronization constr. deletion	0	0						
	0	1						
	0	2						
	1	0	1.1033	0.0799	1.1033	0.0799	1.339	0.1092
	2	0	1.2286	0.1673	1.2052	0.138	1.2962	0.1287
	1	1	1.1327	0.0921	1.1289	0.091	1.3063	0.1092

Table C.19: Evaluation scenario 3: Results on the calculation time in seconds (mean and standard deviation) for the modifications of agent capability variation and agent velocity variation.

Mod.	Constraints in $\mathcal{I}$		INI heuristic		GA-based reoptimization		Conventional GA		Branch-and-price	
	sync.	prec.	mean	SD	mean	SD	mean	SD	mean	SD
Agent capability variation	0	0	0.0009	0.0033	19.676	0.27	19.674	0.174	7.77	2.03
	0	1	0.0027	0.0058	10.797	0.261	10.807	0.281	12.66	11.04
	0	2	0.0019	0.0051	9.837	0.227	9.807	0.234	52.53	62.0
	1	0	0.0028	0.006	10.357	0.271	10.29	0.242	428.74	401.95
	2	0	0.0023	0.0054	8.948	0.27	8.933	0.397	2022.9	1948.39
	1	1	0.0034	0.0058	9.642	0.186	9.585	0.226	531.43	614.29
Agent velocity variation	0	0	0.0025	0.0052	19.542	0.148	19.538	0.14	6.9	2.26
	0	1	0.0019	0.0051	10.749	0.301	10.749	0.233	12.86	14.32
	0	2	0.0031	0.0062	9.803	0.255	9.757	0.215	44.0	53.28
	1	0	0.0037	0.0067	10.332	0.208	10.239	0.2	648.45	839.03
	2	0	0.0016	0.0047	8.938	0.321	8.929	0.358	2638.04	2862.99
	1	1	0.005	0.0073	9.609	0.228	9.573	0.209	473.81	387.26

Table C.20: Evaluation scenario 3: Results on the calculation time in seconds (mean and standard deviation) for the modifications of task position variation and task duration variation.

Mod.	Constraints in $\mathcal{I}$		DIH heuristic		INI heuristic		GA-based reopt. with DIH		GA-based reopt. with INI		Conventional GA		Branch-and-price	
	sync.	prec.	mean	SD	mean	SD	mean	SD	mean	SD	mean	SD	mean	SD
Task position variation	0	0	0.0048	0.0068	0.0034	0.0063	19.526	0.103	19.516	0.12	19.531	0.116	7.96	2.39
	0	1	0.0062	0.007	0.0008	0.0023	10.777	0.294	10.716	0.312	10.786	0.305	19.22	27.8
	0	2	0.0056	0.0057	0.0026	0.0045	9.738	0.226	9.765	0.228	9.778	0.255	34.31	39.18
	1	0	0.0047	0.0072	0.0028	0.006	10.157	0.173	10.219	0.224	10.17	0.207	380.09	427.23
	2	0	0.0065	0.0069	0.0017	0.0035	8.939	0.354	8.934	0.338	8.872	0.297	2159.56	2410.7
	1	1	0.0063	0.0071	0.0016	0.0042	9.534	0.188	9.492	0.183	9.528	0.234	537.72	585.32
Task duration variation	0	0	0.0059	0.0075	0.0014	0.0038	19.506	0.1	19.509	0.107	19.526	0.107	7.51	2.29
	0	1	0.0052	0.0072	0.0023	0.0054	10.817	0.274	10.747	0.251	10.714	0.25	16.13	27.45
	0	2	0.0047	0.0072	0.0028	0.006	9.904	0.328	9.852	0.268	9.851	0.329	30.86	31.23
	1	0	0.0057	0.0072	0.0015	0.0042	10.216	0.177	10.278	0.298	10.27	0.216	476.73	583.61
	2	0	0.0045	0.007	0.0013	0.0042	8.906	0.24	8.907	0.275	8.889	0.315	2075.26	1892.24
	1	1	0.0064	0.0051	0.0025	0.0035	9.571	0.209	9.61	0.216	9.519	0.194	617.66	1247.64

Table C.21: Evaluation scenario 3: Results on the calculation time in seconds (mean and standard deviation) for the modifications of precedence constraint insertion, synchronization constraint insertion, precedence constraint deletion and synchronization constraint deletion.

Mod.	Constraints in $\mathcal{I}$		Reoptimization heuristic		GA-based reoptimization		Conventional GA		Branch-and-price	
	sync.	prec.	mean	SD	mean	SD	mean	SD	mean	SD
Precedence constr. insertion	0	0	0.0125	0.0037	10.809	0.246	10.788	0.298	17.8	21.56
	0	1	0.0137	0.006	9.89	0.201	9.839	0.244	43.76	53.43
	0	2	0.0144	0.0042	9.838	0.259	9.803	0.261	76.88	80.64
	1	0	0.0134	0.0054	9.737	0.234	9.73	0.25	539.66	550.01
	2	0	0.0124	0.0062	9.213	0.174	9.205	0.167	1807.73	1872.11
	1	1	0.0147	0.0048	9.699	0.203	9.678	0.234	733.85	630.01
Synchronization constr. insertion	0	0	0.0169	0.0023	10.076	0.244	9.999	0.21	412.75	375.96
	0	1	0.0176	0.0056	9.523	0.201	9.553	0.212	630.2	954.67
	0	2	0.0167	0.0045	9.684	0.202	9.614	0.204	670.29	616.11
	1	0	0.0142	0.0057	9.027	0.128	8.958	0.138	2025.56	1881.62
	2	0	0.0129	0.0075	8.644	0.116	8.639	0.116	6370.99	5513.63
	1	1	0.015	0.0071	9.203	0.224	9.121	0.196	2697.25	2749.53
Precedence constr. deletion	0	0								
	0	1	0.0022	0.0054	19.691	0.085	19.702	0.1	7.82	2.7
	0	2	0.0022	0.0054	11.502	2.466	11.514	2.485	19.36	29.41
	1	0								
	2	0								
	1	1	0.0037	0.0064	10.144	0.193	10.101	0.216	375.44	520.82
Synchronization constr. deletion	0	0								
	0	1								
	0	2								
	1	0	0.0022	0.0054	19.871	0.109	19.848	0.098	7.88	2.71
	2	0	0.0019	0.0051	15.837	4.605	15.76	4.622	191.21	491.49
	1	1	0.0036	0.0064	10.726	0.315	10.732	0.239	9.68	3.55

Table C.22: Evaluation scenario 3: Results on the Levenshtein distance (mean and standard deviation) for the modifications of task insertion and task deletion.

Mod.	Constraints in $\mathcal{I}$		Reoptimization heuristic		GA-based reoptimization		Conventional GA		Branch-and-price	
	sync.	prec.	mean	SD	mean	SD	mean	SD	mean	SD
Task insertion	0	0	1.0	0.0	1.0	0.0	9.18	2.861	3.72	4.075
	0	1	1.0	0.0	1.0	0.0	9.06	2.983	2.42	3.634
	0	2	1.0	0.0	1.0	0.0	8.58	2.801	2.44	3.281
	1	0	1.0	0.0	1.02	0.14	7.4	2.324	2.86	3.007
	2	0	1.0	0.0	1.0	0.0	6.36	2.152	3.14	3.175
	1	1	1.0	0.0	1.0	0.0	7.3	2.256	2.94	3.036
Task deletion	0	0	1.0	0.0	1.0	0.0	10.08	2.848	4.74	4.476
	0	1	1.0	0.0	1.0	0.0	10.68	3.088	3.82	4.155
	0	2	1.0	0.0	1.0	0.0	10.7	3.055	3.86	4.699
	1	0	1.0	0.0	1.0	0.0	7.96	2.297	4.84	4.032
	2	0	1.0	0.0	1.2	1.4	7.84	2.275	3.98	3.314
	1	1	1.0	0.0	1.0	0.0	9.3	2.715	4.14	4.128

Table C.23: Evaluation scenario 3: Results on the Levenshtein distance (mean and standard deviation) for the modifications of task position variation and task duration variation.

Mod.	Constraints in $\mathcal{I}$		DIH heuristic		INI heuristic		GA-based reopt. with DIH		GA-based reopt. with INI		Conventional GA		Branch-and-price	
	sync.	prec.	mean	SD	mean	SD	mean	SD	mean	SD	mean	SD	mean	SD
Task position variation	0	0	1.2	0.98	0.0	0.0	1.2	0.98	0.0	0.0	10.08	2.682	4.3	4.662
	0	1	1.12	0.993	0.0	0.0	1.38	1.754	0.46	2.08	9.76	3.235	3.54	4.455
	0	2	1.0	1.0	0.0	0.0	1.24	2.074	0.0	0.0	9.7	3.348	4.44	5.543
	1	0	1.24	0.971	0.0	0.0	1.62	1.754	0.18	1.26	8.14	2.332	3.98	4.082
	2	0	1.32	0.947	0.0	0.0	1.72	1.709	0.54	1.91	7.36	2.225	4.26	3.422
	1	1	1.24	0.971	0.0	0.0	1.4	1.562	0.24	1.68	8.16	2.148	3.66	3.037
Task duration variation	0	0	0.04	0.28	0.0	0.0	0.04	0.28	0.0	0.0	9.64	2.719	0.28	1.386
	0	1	0.04	0.28	0.0	0.0	0.04	0.28	0.0	0.0	10.52	3.022	0.0	0.0
	0	2	0.08	0.392	0.0	0.0	0.08	0.392	0.0	0.0	10.32	3.146	0.0	0.0
	1	0	0.56	0.898	0.0	0.0	0.56	0.898	0.0	0.0	8.08	2.505	0.16	0.88
	2	0	0.8	0.98	0.0	0.0	1.12	1.883	0.0	0.0	6.9	2.335	0.5	1.5
	1	1	0.44	0.828	0.0	0.0	0.6	1.562	0.0	0.0	8.4	2.341	0.64	2.313

Table C.24: Evaluation scenario 3: Results on the Levenshtein distance (mean and standard deviation) for the modifications of precedence constraint insertion, synchronization constraint insertion, precedence constraint deletion and synchronization constraint deletion.

Mod.	Constraints in $\mathcal{I}$		Reoptimization heuristic		GA-based reoptimization		Conventional GA		Branch-and-price	
	sync.	prec.	mean	SD	mean	SD	mean	SD	mean	SD
Precedence constr. insertion	0	0	1.16	0.987	1.16	0.987	10.54	2.685	4.38	4.024
	0	1	0.92	0.997	1.26	2.143	10.26	2.999	3.76	5.339
	0	2	1.08	0.997	1.08	0.997	10.44	2.988	4.68	5.931
	1	0	1.2	0.98	1.36	1.572	8.52	2.532	3.28	3.904
	2	0	0.84	0.987	1.1	1.652	8.12	2.188	2.56	3.401
	1	1	0.88	0.993	0.88	0.993	8.12	2.312	2.56	3.054
Synchronization constr. insertion	0	0	1.84	0.543	5.4	4.927	9.88	2.372	8.14	3.784
	0	1	1.92	0.392	3.78	3.874	9.88	2.597	8.04	3.72
	0	2	1.88	0.475	4.54	4.704	9.9	2.563	8.0	3.561
	1	0	1.8	0.6	2.62	2.331	8.14	1.822	6.42	3.269
	2	0	1.88	0.475	2.98	2.565	7.66	2.036	5.48	2.67
	1	1	1.92	0.392	2.34	1.762	7.78	1.847	5.88	3.037
Precedence constr. deletion	0	0								
	0	1	0.0	0.0	0.0	0.0	10.04	3.256	4.76	3.598
	0	2	0.0	0.0	0.0	0.0	9.86	3.169	6.34	6.614
	1	0								
	2	0								
	1	1	0.0	0.0	0.0	0.0	8.08	2.827	3.32	3.997
Synchronization constr. deletion	0	0								
	0	1								
	0	2								
	1	0	0.0	0.0	0.0	0.0	8.38	2.465	7.72	4.099
	2	0	0.0	0.0	2.38	3.888	8.02	2.267	7.94	3.652
	1	1	0.0	0.0	0.66	2.122	8.14	2.482	8.16	3.982

References

Public References

[AAM13] Khalid Abd, Kazem Abhary, and Romeo Marian. Application of Fuzzy Logic to Multi-Objective Scheduling Problems in Robotic Flexible Assembly Cells. *Automation, Control and Intelligent Systems*, 1(3):34–41, 2013. doi:10.11648/j.acis.20130103.11.

[AAS22] Seema A. Alsaidy, Amenah D. Abbood, and Mouayad A. Sahib. Heuristic initialization of PSO task scheduling algorithm in cloud computing. *Journal of King Saud University - Computer and Information Sciences*, 34(6):2370–2382, 2022. doi:10.1016/j.jksuci.2020.11.002.

[ABE09] Giorgio Ausiello, Vincenzo Bonifaci, and Bruno Escoffier. Complexity and Approximation in Reoptimization. In S. Barry Cooper, Andrea Sorbi, and Stuart B. Cooper, editors, *Computability in context*, pages 101–129. Imperial College Pr, London, 2009. doi:10.1142/9781848162778_0004.

[ABLW13] Hanane Allaoua, Sylvie Borne, Lucas Létocart, and Roberto Wolfler Calvo. A matheuristic approach for solving a home health care problem. *Electronic Notes in Discrete Mathematics*, 41:471–478, 2013. doi:10.1016/j.endm.2013.05.127.

[ABS03] Claudia Archetti, Luca Bertazzi, and M. Grazia Speranza. Reoptimizing the traveling salesman problem. *Networks*, 42(3):154–159, 2003. doi:10.1002/net.10091.

[ABS10] Claudia Archetti, Luca Bertazzi, and M. Grazia Speranza. Reoptimizing the 0–1 knapsack problem. *0166-218X*, 158(17):1879–1887, 2010. doi:10.1016/j.dam.2010.08.003.

[AEA+23] Jeffrey O. Agushaka, Absalom E. Ezugwu, Laith Abualigah, Samaher Khalaf Alharbi, and Hamiden Abd El-Wahed Khalifa. Efficient Initialization Methods for Population-Based Metaheuristic Algorithms: A Comparative Study. *Archives of Computational Methods in Engineering*, 30(3):1727–1787, 2023. doi:10.1007/s11831-022-09850-4.

[AEMP09] Giorgio Ausiello, Bruno Escoffier, Jérôme Monnot, and Vangelis Paschos. Reoptimization of minimum and maximum traveling salesman's tours. *Journal of Discrete Algorithms*, 7(4):453–463, 2009. doi:10.1016/j.jda.2008.12.001.

[AH17] Muhammad Usman Arif and Sajjad Haider. An Evolutionary Traveling Salesman Approach for Multi-Robot Task Allocation. In Jaap den van Herik, Ana Paula Rocha, and Joaquim Filipe, editors, *ICAART 2017*, pages 567–574, Setúbal, 2017. SCITEPRESS - Science and Technology Publications Lda. doi:10.5220/0006197305670574.

[AKH03] M. Alighanbari, Y. Kuwata, and J. P. How. Coordination and control of multiple UAVs with timing constraints and loitering. In *Proceedings of the 2003 American Control Conference, ACC*, pages 5311–5316, Piscataway, NJ, 2003. IEEE Service Center. doi:10.1109/ACC.2003.1242572.

[APP+22] Haris Aziz, Arindam Pal, Ali Pourmiri, Fahimeh Ramezani, and Brendan Sims. Task Allocation Using a Team of Robots. *Current Robotics Reports*, 3(4):227–238, 2022. doi:10.1007/s43154-022-00087-4.

[ASGM23] Luke Antonyshyn, Jefferson Silveira, Sidney Givigi, and Joshua Marshall. Multiple Mobile Robot Task and Motion Planning: A Survey. *ACM Computing Surveys*, 55(10):1–35, 2023. doi:10.1145/3564696.

[BBH+08] Davide Bilò, Hans-Joachim Böckenhauer, Juraj Hromkovič, Richard Královič, Tobias Mömke, Peter Widmayer, and Anna Zych. Reoptimization of Steiner Trees. In Joachim Gudmundsson, editor, *Algorithm theory - SWAT 2008*, volume 5124 of *Lecture Notes in Computer Science*, pages 258–269. Springer, Berlin and Heidelberg, 2008. doi:10.1007/978-3-540-69903-3_24.

[BBV08] Roberto Baldacci, Maria Battarra, and Daniele Vigo. Routing a Heterogeneous Fleet of Vehicles. In Bruce L. Golden, S. Raghavan, and Edward Wasil, editors, *The Vehicle Routing Problem: Latest Advances and New Challenges*, volume 43 of *Operations Research/Computer Science Interfaces*, pages 3–27. Springer Science+Business Media, LLC, Boston, MA, 2008. doi:10.1007/978-0-387-77778-8_1.

[BCH19] Noam Buckman, Han-Lim Choi, and Jonathan P. How. Partial Replanning for Decentralized Dynamic Task Allocation. In AIAA, editor, *AIAA Scitech 2019 Forum*, Reston, Virginia, 2019. American Institute of Aeronautics and Astronautics. doi:10.2514/6.2019-0915.

[BF06] Stefan Bertels and Torsten Fahle. A hybrid setup for a hybrid scenario: combining heuristics for the home health care problem. *Computers & Operations Research*, 33(10):2866–2890, 2006. doi:10.1016/j.cor.2005.01.015.

[BFC+16] Manuele Bonaccorsi, Laura Fiorini, Filippo Cavallo, Alessandro Saffiotti, and Paolo Dario. A Cloud Robotics Solution to Improve Social Assistive Robots for Active and Healthy Aging. *International Journal of Social Robotics*, 8(3):393–408, 2016. doi:10.1007/s12369-016-0351-1.

[BFH+06] Hans-Joachim Böckenhauer, Luca Forlizzi, Juraj Hromkovič, Joachim Kneis, Joachim Kupke, Guido Proietti, and Peter Widmayer. Reusing Optimal TSP Solutions for Locally Modified Input Instances. In Gonzalo Navarro, Leopoldo Bertossi, and Yoshiharu Kohayakawa, editors, *Fourth IFIP International Conference on Theoretical Computer Science- TCS 2006*, volume 209 of *IFIP International Federation for Information Processing*, pages 251–270. Springer US, Boston, MA, 2006. doi:10.1007/978-0-387-34735-6_21.

[BH09] Tobias Berg and Harald Hempel. Reoptimization of Traveling Salesperson Problems: Changing Single Edge-Weights. In Adrian-Horia Dediu, Adrian Horia Dediu, Armand Mihai Ionescu, and Carlos Martín Vide, editors, *Language and automata theory and applications*, volume 5457 of *Lecture notes in computer science SL 1 - Theoretical computer science and general issues*, pages 141–151. Springer, Berlin and Heidelberg, 2009. doi:10.1007/978-3-642-00982-2_12.

[BHK13] Mohamed Badreldin, Ahmed Hussein, and Alaa Khamis. A Comparative Study between Optimization and Market-Based Approaches to Multi-Robot Task Allocation. *Advances in Artificial Intelligence*, 2013:1–11, 2013. doi:10.1155/2013/256524.

[BHMW08] Hans-Joachim Böckenhauer, Juraj Hromkovič, Tobias Mömke, and Peter Widmayer. On the Hardness of Reoptimization. In David Hutchison, Takeo Kanade, Josef Kittler, Jon M. Kleinberg, Friedemann Mattern, John C. Mitchell, Moni Naor, Oscar Nierstrasz, C. Pandu Rangan, Bernhard Steffen, Madhu Sudan, Demetri Terzopoulos, Doug Tygar, Moshe Y. Vardi, Gerhard Weikum, Viliam Geffert, Juhani Karhumäki, Alberto Bertoni, Bart Preneel, Pavol Návrat, and Mária Bieliková, editors, *SOFSEM 2008: Theory and Practice of Computer Science*, pages 50–65. Springer, Berlin, Heidelberg, 2008. doi:10.1007/978-3-540-77566-9_5.

[BK10] Hans-Joachim Böckenhauer and Dennis Komm. Reoptimization of the metric deadline TSP. *Journal of Discrete Algorithms*, 8(1):87–100, 2010. doi:10.1016/j.jda.2009.04.001.

[BKAJ15] Patrick Benavidez, Mohan Kumar, Sos Agaian, and Mo Jamshidi. Design of a home multi-robot system for the elderly and disabled. In *2015 10th System of Systems Engineering Conference (SoSE 2015)*, pages 392–397, Piscataway, NJ, 2015. IEEE. doi:10.1109/SYSOSE.2015.7151907.

[BKK09] Hans-Joachim Böckenhauer, Joachim Kneis, and Joachim Kupke. Approximation hardness of deadline-TSP reoptimization. *Theoretical Computer Science*, 410(21-23):2241–2249, 2009. doi:10.1016/j.tcs.2009.02.016.

[BM17] Christian Blum and Max Manfrin. Metaheuristics Network, 02.10.2017. URL: http://www.metaheuristics.org/.

[BNB16] Kyle E. C. Booth, Goldie Nejat, and J. Christopher Beck. A Constraint Programming Approach to Multi-Robot Task Allocation and Scheduling in Retirement Homes. In Michel Rueher, editor, *Principles and practice of constraint programming*, volume 9892 of *LNCS sublibrary. SL 2, Programming and software engineering*, pages 539–555. Springer, Switzerland, 2016. doi:10.1007/978-3-319-44953-1_34.

[BP11] Nicolas Boria and Vangelis T. Paschos. A survey on combinatorial optimization in dynamic environments. *RAIRO - Operations Research*, 45(3):241–294, 2011. doi:10.1051/ro/2011114.

[BPS+20] Kyle Brown, Oriana Peltzer, Martin A. Sehr, Mac Schwager, and Mykel J. Kochenderfer. Optimal Sequential Task Assignment and Path Finding for Multi-Agent Robotic Assembly Planning. *2020 IEEE International Conference on Robotics and Automation (ICRA)*, 2020. doi:10.1109/ICRA40945.2020.

[BQ64] M. L. Balinski and R. E. Quandt. On an Integer Program for a Delivery Problem. *Operations Research*, 12(2):300–304, 1964. doi:10.1287/opre.12.2.300.

[BR02] Michael A. Bender and Dana Ron. Testing properties of directed graphs: acyclicity and connectivity*. *Random Structures & Algorithms*, 20(2):184–205, 2002. doi:10.1002/rsa.10023.

[BR07] David Bredström and Mikael Rönnqvist. A branch and price algorithm for the combined vehicle routing and scheduling problem with synchronization constraints. *NHH Dept. of Finance & Management Science Discussion Paper*, (2007/7), 2007.

[CBH09] Han-Lim Choi, L. Brunet, and J. P. How. Consensus-Based Decentralized Auctions for Robust Task Allocation. *IEEE Transactions on Robotics*, 25(4):912–926, 2009. doi:10.1109/TRO.2009.2022423.

[CFSR23] Chao Cheng, Jun Fu, Hang Su, and Luquan Ren. Recent Advancements in Agriculture Robots: Benefits and Challenges. *Machines*, 11(1):48, 2023. doi:10.3390/machines11010048.

[CGLL23] Hamza Chakraa, François Guérin, Edouard Leclercq, and Dimitri Lefebvre. Optimization techniques for Multi-Robot Task Allocation problems: Review on the state-of-the-art. *Robotics and Autonomous Systems*, 168:104492, 2023. doi:10.1016/j.robot.2023.104492.

[Chr22] Nicos Christofides. Worst-Case Analysis of a New Heuristic for the Travelling Salesman Problem. *Operations Research Forum*, 3(1), 2022. doi: 10.1007/s43069-021-00101-z.

[CK21] Omar Cheikhrouhou and Ines Khoufi. A comprehensive survey on the Multiple Traveling Salesman Problem: Applications, approaches and taxonomy. *Computer Science Review*, 40:100369, 2021. doi:10.1016/j.cosrev.2021.100369.

[CSP+19] Sébastien Cajot, Nils Schüler, Markus Peter, Andreas Koch, and Francois Maréchal. Interactive Optimization With Parallel Coordinates: Exploring Multidimensional Spaces for Decision Support. *Frontiers in ICT*, 5, 2019. doi:10.3389/fict.2018.00032.

[CZDL18] Xinye Chen, Ping Zhang, Guanglong Du, and Fang Li. Ant Colony Optimization Based Memetic Algorithm to Solve Bi-Objective Multiple Traveling Salesmen Problem for Multi-Robot Systems. *IEEE Access*, 6:21745–21757, 2018. doi:10.1109/access.2018.2828499.

[Dav10] Donald Davendra. *Traveling Salesman Problem: Theory and Applications*. IntechOpen, Erscheinungsort nicht ermittelbar, 2010. doi:64896.

[DBW+18] Alessandro Di Nuovo, Frank Broz, Ning Wang, Tony Belpaeme, Angelo Cangelosi, Ray Jones, Raffaele Esposito, Filippo Cavallo, and Paolo Dario. The multi-modal interface of Robot-Era multi-robot services tailored for the elderly. *Intelligent Service Robotics*, 11(1):109–126, 2018. doi:10.1007/s11370-017-0237-6.

[DDD+17] Dr. Shudong Liu, Dr. Ernest Kurniawan, Dr. Peng Hui Tan, Dr. Peng Zhang, Dr. Sumei Sun, Ms. Ye Sarah, Dr. Shudong Liu, Dr. Ernest Kurniawan, Dr. Peng Hui Tan, Dr. Peng Zhang, Dr. Sumei Sun, and Ms. Ye Sarah. Dynamic Scheduling for Heterogeneous Resources with Time Windows and Precedence Relation: 5-8 Nov. 2017. *TENCON 2017 - 2017 IEEE Region 10 Conference*, pages 3045–3050, 2017. doi: 10.1109/TENCON.2017.8228384.

[DL11] Jacques Desrosiers and Marco E. Lübbecke. Branch-Price-and-Cut Algorithms. *Wiley encyclopedia of operations research and management science (EORMS)*, 2011. doi:10.1002/9780470400531.eorms0118.

[DR59] George B. Dantzig and John H. Ramser. The Truck Dispatching Problem. *Management Science*, 6(1):80–91, 1959.

[DR21] Jennifer David and Thorsteinn Rögnvaldsson. Multi-Robot Routing Problem with Min–Max Objective. *Robotics*, 10(4):122, 2021. doi: 10.3390/robotics10040122.

[DRAR15] Sachin Desale, Akhtar Rasool, Sushil Andhale, and Priti Rane. Heuristic and meta-heuristic algorithms and their relevance to the real world: a survey. *International Journal of Computer Engineering in Research Trends*, 2(5):296–304, 2015.

[DRL11] Anders Dohn, Matias Sevel Rasmussen, and Jesper Larsen. The vehicle routing problem with time windows and temporal dependencies. *Networks*, 58(4):273–289, 2011. doi:10.1002/net.20472.

[DS15] Rina A. Doherty and Paul Sorenson. Keeping Users in the Flow: Mapping System Responsiveness with User Experience. *Procedia Manufacturing*, 3:4384–4391, 2015. doi:10.1016/j.promfg.2015.07.436.

[EA20] Raafat Elshaer and Hadeer Awad. A taxonomic review of metaheuristic algorithms for solving the vehicle routing problem and its variants. *Computers & Industrial Engineering*, 140:106242, 2020. doi:10.1016/j.cie.2019.106242.

[EAR23] Said Elatar, Karim Abouelmehdi, and Mohammed Essaid Riffi. The vehicle routing problem in the last decade: variants, taxonomy and metaheuristics. *Procedia Computer Science*, 220:398–404, 2023. doi:10.1016/j.procs.2023.03.051.

[EM20] Zahra Entezari and Masoud Mahootchi. Developing a mathematical model for staff routing and scheduling in home health care industries: Genetic Algorithm based solution scheme. *Scientia Iranica*, 0(0):3692–3718, 2020. doi:10.24200/sci.2020.54116.3600.

[EMN⁺20] Yousef Emam, Siddharth Mayya, Gennaro Notomista, Addison Bohannon, and Magnus Egerstedt. Adaptive Task Allocation for Heterogeneous Multi-Robot Teams with Evolving and Unknown Robot Capabilities. In *2020 IEEE International Conference on Robotics and Automation (ICRA)*, pages 7719–7725. IEEE, 2020. doi:10.1109/icra40945.2020.9197283.

[EMP09] Bruno Escoffier, Martin Milanič, and Vangelis Paschos. Simple and Fast Reoptimizations for the Steiner Tree Problem. *Algorithmic Operations Research*, 4(2):86–94, 2009.

[ES15] Agoston E. Eiben and James E. Smith. *Introduction to evolutionary computing*. Natural computing series. Springer, Berlin and Heidelberg and New York and Dordrecht and London, second edition, 2015. doi:10.1007/978-3-662-44874-8.

[ES22] H. A. Eiselt and Carl-Louis Sandblom. *Operations Research*. Springer International Publishing, Cham, 2022. doi:10.1007/978-3-030-97162-5.

[FIN17] Alessandro Farinelli, Luca Iocchi, and Daniele Nardi. Distributed online dynamic task assignment for multi-robot patrolling. *Autonomous Robots*, 41(6):1321–1345, 2017. doi:10.1007/s10514-016-9579-8.

[GC22] Payam Ghassemi and Souma Chowdhury. Multi-robot task allocation in disaster response: Addressing dynamic tasks with deadlines and robots with range and payload constraints. *Robotics and Autonomous Systems*, 147:103905, 2022. doi:10.1016/j.robot.2021.103905.

[Ger03] Brian Paul Gerkey. *On Multi-Robot Task Allocation*. Dissertation, University of Southern California, Los Angeles, 2003.

[Gin17] Maria Gini. Multi-Robot Allocation of Tasks with Temporal and Ordering Constraints. In AAAI, editor, *Proceedings of the AAAI Conference on Artificial Intelligence*, volume 31, pages 4863–4869, 2017.

[GM04] Brian P. Gerkey and Maja J. Matarić. A formal analysis and taxonomy of task allocation in multi-robot systems. *The International Journal of Robotics Research*, 23(9):939–954, 2004. doi:10.1177/0278364904045564.

[GM12] Avinash Gautam and Sudeept Mohan. A review of research in multi-robot systems. In *2012 IEEE 7th International Conference on Industrial and Information Systems (ICIIS 2012)*, pages 1–5, Piscataway NJ, 2012. IEEE. doi:10.1109/ICIInfS.2012.6304778.

[GMN77] B. L. Golden, T. L. Magnanti, and H. Q. Nguyen. Implementing vehicle routing algorithms. *Networks*, 7(2):113–148, 1977. doi:10.1002/net.3230070203.

[GP10] Michel Gendreau and Jean-Yves Potvin. *Handbook of Metaheuristics*, volume 146. Springer US, Boston, MA, 2010. doi:10.1007/978-1-4419-1665-5.

[GWS18] Matthew C. Gombolay, Ronald J. Wilcox, and Julie A. Shah. Fast Scheduling of Robot Teams Performing Tasks With Temporospatial Constraints. *IEEE Transactions on Robotics*, 34(1):220–239, 2018. doi:10.1109/TRO.2018.2795034.

[GXCW20] Miao Guo, Bin Xin, Jie Chen, and Yipeng Wang. Multi-agent coalition formation by an efficient genetic algorithm with heuristic initialization and repair strategy. *Swarm and Evolutionary Computation*, 55:100686, 2020. doi:10.1016/j.swevo.2020.100686.

[HGQ+12] Simon Hamel, Jonathan Gaudreault, Claude-Guy Quimper, Mathieu Bouchard, and Philippe Marier. Human-machine interaction for real-time linear optimization. In *IEEE International Conference on Systems, Man, and Cybernetics (SMC), 2012*, pages 673–680, Piscataway, NJ, 2012. IEEE. doi:10.1109/ICSMC.2012.6377804.

[HLP14] Syrine Roufaida Ait Haddadene, Nacima Labadie, and Caroline Prodhon. The VRP with time windows, synchronization and precedence constraints: Application in home health care sector. In *International Conference on Modeling, Optimization and SIMulation - MOSIM14 -*, 2014.

[Hol] HoLLiECares: Multifunktionaler Serviceroboter zur Unterstützung professioneller Pflege in Krankenhäusern. URL: https://www.interaktive-technologien.de/projekte/holliecares.

[HP13] Bradford Heap and Maurice Pagnucco. Repeated Sequential Single-Cluster Auctions with Dynamic Tasks for Multi-Robot Task Allocation with Pickup and Delivery. In Matthias Klusch, Matthias Thimm, and Marcin Paprzycki, editors, *Multiagent System Technologies*, volume 8076 of *Lecture Notes in Computer Science Ser*, pages 87–100. Springer Berlin Heidelberg, Berlin/Heidelberg, 2013. doi:10.1007/978-3-642-40776-5_10.

[HRJL08] Anders Dohn Hansen, Matias Sevel Rasmussen, Tor Justesen, and Jesper Larsen. The Home Care Crew Scheduling Problem. *Conference Proceedings of the 1st International Conference on Applied Operational Research*, 2008.

[HYBB20] Samira Hayat, Evşen Yanmaz, Christian Bettstetter, and Timothy X. Brown. Multi-objective drone path planning for search and rescue with quality-of-service requirements. *Autonomous Robots*, 44(7):1183–1198, 2020. doi:10.1007/s10514-020-09926-9.

[IBM] IBM. IBM ILOG CPLEX Optimization Studio. URL: https://www.ibm.com/de-de/products/ilog-cplex-optimization-studio.

[ITV14] Stefan Irnich, Paolo Toth, and Daniele Vigo. The Family of Vehicle Routing Problems. In Paolo Toth and Daniele Vigo, editors, *Vehicle Routing*, MOS-SIAM series on optimization, pages 1–33. Society for Industrial and Applied Mathematics and Mathematical Optimization Society, Philadelphia, PA, 2014. doi:10.1137/1.9781611973594.ch1.

[JDS11] E. Gil Jones, M. Bernardine Dias, and Anthony Stentz. Time-extended multi-robot coordination for domains with intra-path constraints. *Autonomous Robots*, 30(1):41–56, 2011. doi:10.1007/s10514-010-9202-3.

[Jun13] Dieter Jungnickel. *Graphs, networks and algorithms*, volume volume 5 of *Algorithms and computation in mathematics*. Springer, Berlin and Heidelberg and New York and Dordrecht and London, fourth edition, 2013. doi:10.1007/978-3-642-32278-5.

[Kah62] A. B. Kahn. Topological sorting of large networks. *Communications of the ACM*, 5(11):558–562, 1962. doi:10.1145/368996.369025.

[KHE15] Alaa Khamis, Ahmed Hussein, and Ahmed Elmogy. Multi-robot Task Allocation: A Review of the State-of-the-Art. In Anis Koubâa and J.Ramiro Martínez-de Dios, editors, *Cooperative Robots and Sensor Networks 2015*, volume 604 of *Studies in Computational Intelligence*, pages 31–51. Springer International Publishing and Imprint and Springer, Cham, 2015. doi:10.1007/978-3-319-18299-5_2.

[KKB+12] G. Ayorkor Korsah, Balajee Kannan, Brett Browning, Anthony Stentz, and M. Bernardine Dias. xBots: An approach to generating and executing optimal multi-robot plans with cross-schedule dependencies. In IEEE, editor, *2012 IEEE International Conference on Robotics and Automation (ICRA 2012)*, pages 115–122, Piscataway, NJ, 2012. IEEE. doi: 10.1109/ICRA.2012.6225234.

[KLB09] Yannik Kergosien, Christophe Lenté, and Jean-Charles Billaut. Home health care problem: An extended multiple Traveling Salesman Problem. *Multidisciplinary International Conference on Scheduling : Theory and Applications (MISTA 2009)*, pages 85–92, 2009.

[Kor11] G. Ayorkor Korsah. *Exploring Bounded Optimal Coordination for Heterogeneous Teams with Cross-Schedule Dependencies.* Dissertation, Carnegie Mellon University, Pittsburgh, USA, 2011.

[KPDH12] Attila A. Kovacs, Sophie N. Parragh, Karl F. Doerner, and Richard F. Hartl. Adaptive large neighborhood search for service technician routing and scheduling problems. *Journal of Scheduling*, 15(5):579–600, 2012. doi:10.1007/s10951-011-0246-9.

[Kra17] Oliver Kramer, editor. *Genetic algorithm essentials*, volume volume 679 of *Studies in Computational Intelligence*. Springer, Cham, 2017. doi:10.1007/978-3-319-52156-5.

[KSD13] G. Ayorkor Korsah, Anthony Stentz, and M. Bernardine Dias. A comprehensive taxonomy for multi-robot task allocation. *The International Journal of Robotics Research*, 32(12):1495–1512, 2013. doi:10.1177/0278364913496484.

[KYCL18] Kam Fai Elvis Tsang, Yuqing Ni, Cheuk Fung Raphael Wong, and Ling Shi. *2018 15th International Conference on Control, Automation, Robotics and Vision (ICARCV): 18-21 Nov. 2018.* IEEE, Piscataway, NJ, 2018.

[LCS11] Lingzhi Luo, Nilanjan Chakraborty, and Katia Sycara. Multi-robot assignment algorithm for tasks with set precedence constraints. In IEEE, editor, *2011 IEEE International Conference on Robotics and Automation*, pages 2526–2533. IEEE, 2011. doi:10.1109/icra.2011.5980500.

[LCS12] Lingzhi Luo, Nilanjan Chakraborty, and Katia Sycara. Competitive analysis of repeated greedy auction algorithm for online multi-robot task assignment. In IEEE, editor, *2012 IEEE International Conference on Robotics and Automation*, pages 4792–4799. IEEE, 2012. doi:10.1109/icra.2012.6225195.

[LCS13] Lingzhi Luo, Nilanjan Chakraborty, and Katia Sycara. Distributed algorithm design for multi-robot generalized task assignment problem. In IEEE, editor, *2013 IEEE/RSJ International Conference on Intelligent Robots and Systems*, pages 4765–4771. IEEE, 2013. doi:10.1109/iros.2013.6697043.

[LCS15a] Lingzhi Luo, Nilanjan Chakraborty, and Katia Sycara. Distributed Algorithms for Multirobot Task Assignment With Task Deadline Constraints. *IEEE Transactions on Automation Science and Engineering*, 12(3):876–888, 2015. doi:10.1109/TASE.2015.2438032.

[Lee18] Dong-Hyun Lee. Resource-based task allocation for multi-robot systems. *Robotics and Autonomous Systems*, 103:151–161, 2018. doi:10.1016/j.robot.2018.02.016.

[LHD12] David Landén, Fredrik Heintz, and Patrick Doherty. Complex Task Allocation in Mixed-Initiative Delegation: A UAV Case Study. In Nirmit Desai, Alan Liu, and Michael Winikoff, editors, *Principles and Practice of Multi-Agent Systems*, volume 7057 of *Lecture Notes in Computer Science*, pages 288–303. Springer Berlin Heidelberg, Berlin, Heidelberg, 2012. doi:10.1007/978-3-642-25920-3_20.

[Lis13] Bertrand Lisbach. *Linguistic Identity Matching*. Springer Fachmedien Wiesbaden GmbH, Wiesbaden, first edition, 2013.

[LIW+23] Dongwei Li, Joshua Ignatius, Dujuan Wang, Yunqiang Yin, and T.C.E. Cheng. A branch–and–price–and–cut algorithm for the truck–drone routing problem with simultaneously delivery and pickup. *Naval Research Logistics Quarterly*, 2023. doi:10.1002/nav.22151.

[LK12] Chun Liu and Andreas Kroll. A Centralized Multi-Robot Task Allocation for Industrial Plant Inspection by Using A* and Genetic Algorithms. In Leszek Rutkowski, Marcin Korytkowski, Rafał Scherer, Ryszard Tadeusiewicz, Lotfi A. Zadeh, and Jacek M. Zurada, editors, *International Conference on Artificial Intelligence and Soft Computing*, pages 466–474. Springer, Berlin, Heidelberg, 2012. doi:10.1007/978-3-642-29350-4_56.

[LK15] Chun Liu and Andreas Kroll. Memetic algorithms for optimal task allocation in multi-robot systems for inspection problems with cooperative tasks. *Soft Computing*, 19(3):567–584, 2015. doi:10.1007/s00500-014-1274-0.

[LLW+18] Ning Li, Minglong Li, Yanzhen Wang, Da Huang, and Wei Yi. Fault-Tolerant and Self-Adaptive Market-Based Coordination Using Hoplites Framework for Multi-Robot Patrolling Tasks. In *2018 IEEE International Conference on Real-Time Computing and Robotics*, pages 514–519, Piscataway, NJ, 2018. IEEE. doi:10.1109/RCAR.2018.8621723.

[LLY20] Qian Li, San-Yang Liu, and Xin-She Yang. Influence of initialization on the performance of metaheuristic optimizers. *Applied Soft Computing*, 91:106193, 2020. doi:10.1016/j.asoc.2020.106193.

[Lüb11] Marco E. Lübbecke. Column Generation. In James J. Cochran, Louis A. Cox, Pinar Keskinocak, Jeffrey P. Kharoufeh, and J. Cole Smith, editors, *Wiley Encyclopedia of Operations Research and Management Science*. Wiley, 2011. doi:10.1002/9780470400531.eorms0158.

[LV] Christian Lengenfelder and Michael Voigt. Multifunktionaler Serviceroboter zur Unterstützung professioneller Pflege in Krankenhäusern – HoLLiECares: Teilvorhaben: Intuitive Mensch-Roboter-Interaktion für den Einsatz von Servicerobotik im Krankenhaus: Schlussbericht. URL: https://publica-rest.fraunhofer.de/server/api/core/bitstreams/08261c32-2147-44c2-af90-1ac2a385f57f/content.

[Man13] Atsushi Mano. Smart life designed by robotics and NEDO's outlook on robot technologies. In *2013 IEEE Workshop on Advanced Robotics and its Social Impacts*. IEEE, 2013. doi:10.1109/arso.2013.6705519.

[MBG15] Hakim Mitiche, Dalila Boughaci, and Maria Gini. Efficient heuristics for a time-extended multi-robot task allocation problem. In *2015 First International Conference on New Technologies of Information and Communication (NTIC 2015)*, pages 1–6, Piscataway, NJ, 2015. IEEE. doi:10.1109/NTIC.2015.7368756.

[MFGC21] J. G. Martin, J. R. D. Frejo, R. A. García, and E. F. Camacho. Multi-robot task allocation problem with multiple nonlinear criteria using branch and bound and genetic algorithms. *Intelligent Service Robotics*, 14(5):707–727, 2021. doi:10.1007/s11370-021-00393-4.

[MGDVF21] Adyson M. Maia, Yacine Ghamri-Doudane, Dario Vieira, and Miguel Franklin de Castro. An improved multi-objective genetic algorithm with heuristic initialization for service placement and load distribution in edge computing. *Computer Networks*, 194:108146, 2021. doi:10.1016/j.comnet.2021.108146.

[Mil68] Robert B. Miller. Response time in man-computer conversational transactions. In Association for Computing Machinery, editor, *Proceedings of the December 9-11, 1968, fall joint computer conference, part I on - AFIPS '68*, pages 267–277, New York, New York, USA, 1968. ACM Press. doi:10.1145/1476589.1476628.

[MKF+15] David Meignan, Sigrid Knust, Jean-Marc Frayret, Gilles Pesant, and Nicolas Gaud. A Review and Taxonomy of Interactive Optimization Methods in Operations Research. *ACM Transactions on Interactive Intelligent Systems*, 5(3):1–43, 2015. doi:10.1145/2808234.

[MM06] Alejandro R. Mosteo and Luis Montano. Simulated annealing for multi-robot hierarchical task allocation with flexible constraints and objective functions. In *Workshop on network robot systems: toward intelligent robotic systems integrated with environments. int. conf. on intelligent robots and systems*, 2006.

[Mon15] Jerome Monnot. A note on the traveling salesman reoptimization problem under vertex insertion. *Information Processing Letters*, 115(3):435–438, 2015. doi:10.1016/j.ipl.2014.11.003.

[MRD07] Waqar Malik, Sivakumar Rathinam, and Swaroop Darbha. An approximation algorithm for a symmetric Generalized Multiple Depot, Multiple Travelling Salesman Problem. *Operations Research Letters*, 35(6):747–753, 2007. doi:10.1016/j.orl.2007.02.001.

[MSF+14] I. Maurtua, L. Susperregi, A. Fernández, C. Tubío, C. Perez, J. Rodríguez, T. Felsch, and M. Ghrissi. MAINBOT – Mobile Robots for Inspection and Maintenance in Extensive Industrial Plants. *Energy Procedia*, 49:1810–1819, 2014. doi:10.1016/j.egypro.2014.03.192.

[NC12] Ferrante Neri and Carlos Cotta. Memetic algorithms and memetic computing optimization: A literature review. *Swarm and Evolutionary Computation*, 2:1–14, 2012. doi:10.1016/j.swevo.2011.11.003.

[NG15] Ernesto Nunes and Maria Gini. Multi-Robot Auctions for Allocation of Tasks with Temporal Constraints. *Proceedings of the AAAI Conference on Artificial Intelligence*, 29(1), 2015. doi:10.1609/aaai.v29i1.9440.

[NMMG17] Ernesto Nunes, Marie Manner, Hakim Mitiche, and Maria Gini. A taxonomy for task allocation problems with temporal and ordering constraints. *Robotics and Autonomous Systems*, 90:55–70, 2017. doi:10.1016/j.robot.2016.10.008.

[NW06] Jorge Nocedal and Stephen J. Wright. *Numerical optimization*. Springer series in operation research and financial engineering. Springer, New York, NY, second edition, 2006.

[OL96] Ibrahim H. Osman and Gilbert Laporte. Metaheuristics: A bibliography. *Annals of Operations Research*, 63(5):511–623, 1996. doi:10.1007/BF02125421.

[Oxf] How Robots Change the World: What Automation Really Means for Jobs and Productivity.

[OXM+21] Brenner Humberto Ojeda Rios, Eduardo C. Xavier, Flávio K. Miyazawa, Pedro Amorim, Eduardo Curcio, and Maria João Santos. Recent dynamic vehicle routing problems: A survey. *Computers & Industrial Engineering*, 160:107604, 2021. doi:10.1016/j.cie.2021.107604.

[PAN23] Ping-Qi PAN. *Linear Programming Computation*. Springer Nature Singapore and Imprint Springer, Singapore, second edition, 2023. doi:10.1007/978-981-19-0147-8.

[Par98] L. E. Parker. ALLIANCE: an architecture for fault tolerant multirobot cooperation. *IEEE Transactions on Robotics and Automation*, 14(2):220–240, 1998. doi:10.1109/70.681242.

[Pea84] Judea Pearl. *Heuristics: Intelligent search strategies for computer problem solving*. The Addison-Wesley series in artifical intelligence. Addison-Wesley Publishing Company, Reading, Massachusetts and Menlo Park, California and London and Amsterdam and Don Mills, Ontario and Sydney, 1984.

[PGGM13] Victor Pillac, Michel Gendreau, Christelle Guéret, and Andrés L. Medaglia. A review of dynamic vehicle routing problems. *European Journal of Operational Research*, 225(1):1–11, 2013. doi:10.1016/j.ejor.2012.08.015.

[PGM13] V. Pillac, C. Guéret, and A. L. Medaglia. A parallel matheuristic for the technician routing and scheduling problem. *Optimization Letters*, 7(7):1525–1535, 2013. doi:10.1007/s11590-012-0567-4.

[PR19] David Pisinger and Stefan Ropke. Large Neighborhood Search. In Michel Gendreau and Jean-Yves Potvin, editors, *Handbook of Metaheuristics*, volume 272 of *International Series in Operations Research & Management Science*, pages 99–127. Springer International Publishing, Cham, 2019. doi:10.1007/978-3-319-91086-4_4.

[PRS16] Lynne E. Parker, Daniela Rus, and Gaurav S. Sukhatme. Multiple Mobile Robot Systems. In Bruno Siciliano and Oussama Khatib, editors, *Springer handbook of robotics*, Springer Handbooks, pages 1335–1384. Springer, Berlin, 2016. doi:10.1007/978-3-319-32552-1_53.

[PRSB18] K. Padmanabhan Panchu, M. Rajmohan, R. Sundar, and R. Baskaran. Multi-objective Optimisation of Multi-robot Task Allocation with Precedence Constraints. *Defence Science Journal*, 68(2):175, 2018. doi:10.14429/dsj.68.11187.

[RAT19] Yara Rizk, Mariette Awad, and Edward W. Tunstel. Cooperative Heterogeneous Multi-Robot Systems; A Survey. *ACM Computing Surveys*, 52(2):1–31, 2019. doi:10.1145/3303848.

[RDW+00] Reid Simmons, David Apfelbaum, Wolfram Burgard, Dieter Fox, Mark Moors, Sebastian Thrun, and Hakan Younes. Coordination for Multi-Robot Exploration and Mapping. In AAAI, editor, *Proceedings of the AAAI Conference on Artificial Intelligence (IAAI-2000)*, pages 852–858, Menlo Park, Calif. and Cambridge, Mass. and London, 2000. AAAI Press and MIT Press.

[RJDL12] Matias Sevel Rasmussen, Tor Justesen, Anders Dohn, and Jesper Larsen. The Home Care Crew Scheduling Problem: Preference-based visit clustering and temporal dependencies. *European Journal of Operational Research*, 219(3):598–610, 2012. doi:10.1016/j.ejor.2011.10.048.

[Rot22] Simon Rothfuß. *Human-Machine Cooperative Decision Making*. Dissertation, Karlsruhe Institute of Technology, Karlsruhe, 20022.

[Sal17] Saïd Salhi. *Heuristic search: The emerging science of problem solving*. Palgrave Macmillan, Cham, Switzerland, 2017. doi:10.1007/978-3-319-49355-8.

[Seo08] Steven Seow. *Designing and Engineering Time: The Psychology of Time Perception in Software*. Pearson Education, Limited, Hoboken, 2008.

[Shn87] Ben Shneiderman. Designing the user interface strategies for effective human-computer interaction. *ACM SIGBIO Newsletter*, 9(1):6, 1987. doi:10.1145/25065.950626.

[SK16a] Amit Shukla and Hamad Karki. Application of robotics in onshore oil and gas industry—A review Part I. *Robotics and Autonomous Systems*, 75:490–507, 2016. doi:10.1016/j.robot.2015.09.012.

[SKRJ20] N. Seenu, R. M. Kuppan Chetty, M. M. Ramya, and M. N. Janardhanan. Review on state-of-the-art dynamic task allocation strategies for multiple-robot systems. *Industrial Robot: the international journal of robotics research and application*, 47(6):929–942, 2020. doi:10.1108/IR-04-2020-0073.

[SSPÖ15] Eric Schneider, Elizabeth I. Sklar, Simon Parsons, and A. Tuna Özgelen. Auction-Based Task Allocation for Multi-robot Teams in Dynamic Environments. In Clare Dixon and Karl Tuyls, editors, *Towards autonomous robotic systems*, volume 9287 of *Lecture notes in computer science Lecture notes in artificial intelligence*, pages 246–257. Springer, Cham, 2015. doi:10.1007/978-3-319-22416-9_29.

[STBE09] S. Sariel-Talay, T. R. Balch, and N. Erdogan. Multiple Traveling Robot Problem: A Solution Based on Dynamic Task Selection and Robust Execution. *IEEE/ASME Transactions on Mechatronics*, 14(2):198–206, 2009. doi:10.1109/TMECH.2009.2014157.

[SVJ23] Malek Sarhani, Stefan Voß, and Raka Jovanovic. Initialization of metaheuristics: comprehensive review, critical analysis, and research directions. *International Transactions in Operational Research*, 30(6):3361–3397, 2023. doi:10.1111/itor.13237.

[SW12] Robert Sedgewick and Kevin Wayne. *Algorithms*. Addison-Wesley, Upper Saddle River, NJ, fourth edition, 2012.

[Tan15] Jun Tanimoto. *Fundamentals of Evolutionary Game Theory and its Applications*, volume 6. Springer Japan, Tokyo, 2015. doi:10.1007/978-4-431-54962-8.

[TLM19] Salma Hadj Taieb, Taicir Loukil, and Abderrahman El Mhamedi. Home (Health)-Care Routing and Scheduling Problem. In IEEE, editor, *International Conference on Logistics and Supply Chain Management*, pages 1–6, Piscataway, NJ, 2019. IEEE. doi:10.1109/LOGISTIQUA.2019.8907292.

[TV02] Paolo Toth and Daniele Vigo. *The vehicle routing problem*. SIAM monographs on discrete mathematics and applications. Society for Industrial and Applied Mathematics, Philadelphia, 2002.

[VR21] Janardan Kumar Verma and Virender Ranga. Multi-Robot Coordination Analysis, Taxonomy, Challenges and Future Scope. *Journal of Intelligent & Robotic Systems*, 102(1):10, 2021. doi:10.1007/s10846-021-01378-2.

[WC21] Hanfu Wang and Weidong Chen. Task scheduling for transport and pick robots in logistics: a comparative study on constructive heuristics. *Autonomous Intelligent Systems*, 1(1), 2021. doi:10.1007/s43684-021-00017-9.

[WC22] Hanfu Wang and Weidong Chen. Simulated Annealing Algorithms for the Heterogeneous Robots Task Scheduling Problem in Heterogeneous Robotic Order Fulfillment Systems. In Marcelo H. Ang Jr, Hajime Asama, Wei Lin, and Shaohui Foong, editors, *Intelligent Autonomous Systems 16*, volume 412 of *Lecture Notes in Networks and Systems*, pages 276–287. Springer International Publishing and Imprint Springer, Cham, 2022. doi:10.1007/978-3-030-95892-3_21.

[WJC20] Changyun Wei, Ze Ji, and Boliang Cai. Particle Swarm Optimization for Cooperative Multi-Robot Task Allocation: A Multi-Objective Approach. *IEEE Robotics and Automation Letters*, 5(2):2530–2537, 2020. doi:10.1109/lra.2020.2972894.

[XXR11] Zhou Xu, Liang Xu, and Brian Rodrigues. An analysis of the extended Christofides heuristic for the -depot TSP. *Operations Research Letters*, 39(3):218–223, 2011. doi:10.1016/j.orl.2011.03.002.

[Yak16] Ertan Yakici. Solving location and routing problem for UAVs. *Computers & Industrial Engineering*, 102:294–301, 2016. doi:10.1016/j.cie.2016.10.029.

[YJC13] Zhi Yan, Nicolas Jouandeau, and Arab Ali Cherif. A Survey and Analysis of Multi-Robot Coordination. *International Journal of Advanced Robotic Systems*, 10(12), 2013. doi:10.5772/57313.

[YJRD11] S. Yadlapalli, Jungyun Bae, S. Rathinam, and S. Darbha. Approximation algorithms for a heterogeneous Multiple Depot Hamiltonian Path Problem. In IEEE, editor, *Proceedings of the 2011 American Control Conference*, pages 2789–2794. IEEE, 2011. doi:10.1109/acc.2011.5991534.

[YRD10] Sai Yadlapalli, Sivakumar Rathinam, and Swaroop Darbha. 3-Approximation algorithm for a two depot, heterogeneous traveling salesman problem. *Optimization Letters*, 6:141–152, 2010. doi:10.1007/s11590-010-0256-0.

[YZM+21] Shujun Yang, Yichuan Zhang, Lianbo Ma, Yan Song, Ping Zhou, Gang Shi, and Hanning Chen. A Novel Maximin-Based Multi-Objective Evolutionary Algorithm Using One-by-One Update Scheme for Multi-Robot Scheduling Optimization. *IEEE Access*, 9:121316–121328, 2021. doi:10.1109/ACCESS.2021.3105102.

[ZBB10] Günther Zäpfel, Roland Braune, and Michael Bögl. *Metaheuristic Search Concepts*. Springer Berlin Heidelberg, Berlin, Heidelberg, 2010. doi:10.1007/978-3-642-11343-7.

[Zlo06] Robert Zlot. *An Auction-Based Approach to Complex Task Allocation for Multirobot Teams*. Dissertation, Carnagie Mellon University, Pittsburgh, USA, 2006.

[ZP13c] Yu Zhang and Lynne E. Parker. Multi-Robot Task Scheduling. *IEEE International Conference on Robotics and Automation (ICRA)*, pages 2992–2998, 2013. doi:10.1109/ICRA.2013.6630992.

[ZS06] Robert Zlot and Anthony Stentz. Market-based Multirobot Coordination for Complex Tasks. *The International Journal of Robotics Research*, 25(1):73–101, 2006. doi:10.1177/0278364906061160.

[ZS17] Emilio Zamorano and Raik Stolletz. Branch-and-price approaches for the Multiperiod Technician Routing and Scheduling Problem. *European Journal of Operational Research*, 257(1):55–68, 2017. doi:10.1016/j.ejor.2016.06.058.

[ZTY17] Zhanxia Zhu, Biwei Tang, and Jianping Yuan. Multirobot task allocation based on an improved particle swarm optimization approach. *International Journal of Advanced Robotic Systems*, 14(3):172988141771031, 2017. doi:10.1177/1729881417710312.

[ZXS14] Yan Zheng, Yujie Xiao, and Yoonho Seo. A tabu search algorithm for simultaneous machine/AGV scheduling problem. *International Journal of Production Research*, 52(19):5748–5763, 2014. doi:10.1080/00207543.2014.910628.

Own Publications and Conference Contributions

[BKH+24] Esther Bischoff, Saskia Kohn, Daniela Hahn, Christian Braun, Simon Rothfuß, and Sören Hohmann. Heuristic reoptimization of time–extended multi–robot task allocation problems. *Networks*, 2024. doi:10.1002/net.22217.

[BMIH20] Esther Bischoff, Fabian Meyer, Jairo Inga, and Soren Hohmann. Multi-Robot Task Allocation and Scheduling Considering Cooperative Tasks and Precedence Constraints. In IEEE, editor, *2020 IEEE International Conference on Systems, Man, and Cybernetics (SMC)*. IEEE, 2020. doi:10.1109/smc42975.2020.9283215.

[BTIH21] Esther Bischoff, Jonas Teufel, Jairo Inga, and Soren Hohmann. Towards interactive coordination of heterogeneous robotic teams – Introduction of a reoptimization framework. In IEEE, editor, *2021 IEEE International Conference on Systems, Man, and Cybernetics (SMC)*. IEEE, 2021. doi:10.1109/smc52423.2021.9659002.

[BWRH24] Esther Bischoff, Robin Wöran, Simon Rothfuß, and Sören Hohmann. Multi-robot task allocation with complex dependencies – a branch-and-price approach. In Guido Voigt, Malte Fliedner, Knut Haase, Wolfgang Brüggemann, Kai Hoberg, and Joern Meissner, editors, *Operations Research Proceedings 2023*, Lecture Notes in Operations Research. Springer International Publishing and Imprint Springer, Cham, 2024.

[SMB+20] Martin J. Schuster, Marcus G. Muller, Sebastian G. Brunner, Hannah Lehner, Peter Lehner, Ryo Sakagami, Andreas Domel, Lukas Meyer, Bernhard Vodermayer, Riccardo Giubilato, Mallikarjuna Vayugundla, Josef Reill, Florian Steidle, Ingo von Bargen, Kristin Bussmann, Rico Belder, Philipp Lutz, Wolfgang Sturzl, Michal Smisek, Moritz Maier, Samantha Stoneman, Andre Fonseca Prince, Bernhard Rebele, Maximilian Durner, Emanuel Staudinger, Siwei Zhang, Robert Pohlmann, Esther Bischoff, Christian Braun, Susanne Schroder, Enrico Dietz, Sven Frohmann, Anko Borner, Heinz-Wilhelm Hubers, Bernard Foing, Rudolph Triebel, Alin O. Albu-Schaffer, and Armin Wedler. The ARCHES Space-Analogue Demonstration Mission: Towards Heterogeneous Teams of Autonomous Robots for Collaborative Scientific Sampling in Planetary Exploration. *IEEE Robotics and Automation Letters*, 5(4):5315–5322, 2020. doi:10.1109/LRA.2020.3007468.

[WMS+21] Armin Wedler, Marcus Gerhard Müller, Martin Schuster, Maximilian Durner, Sebastian Brunner, Peter Lehner, Hannah Lehner, Andreas Dömel, Mallikarjuna Vayugundla, Florian Steidle, Ryo Sakagami, Lukas Meyer, Michal Smisek, Wolfgang Stürzl, Nicole Schmitz, Bernhard Vodermayer, Andre Fonseca Prince, Emanuel Staudinger, Matthias Hellerer, Roy Lichtenheldt, Enrico Dietz, Christian Braun, Bernhard Rebele, Wout Boerdijk,

Riccardo Giubilato, Josef Reill, Moritz Kuhne, Jongseok Lee, Alejandro Fontan Villacampa, Ingo Bargen, Susanne Schröder, Sven Frohmann, Fabian Seel, Rudolph Triebel, Neal Yi-Sheng Lii, Esther Bischoff, Sean Kille, Kjetil Wormnes, Aaron Pereira, William Carey, Angelo P. Rossi, Laurenz Thomsen, Thorsten Graber, Thomas Krüger, Peter Kyr, Anko Börner, Kristin Bussmann, Gerhard Paar, Arnold Bauer, Stefan Völk, Andreas Kimpe, Heike Rauer, Heinz-Wilhelm Hübers, Johann Bals, Sören Hohmann, Tamim Asfour, Bernard Foing, and Alin Albu-Schäffer. Preliminary Results for the Multi-Robot, Multi-Partner, Multi-Mission, Planetary Exploration Analogue Campaign on Mount Etna. In IAC, editor, *72nd International Astronautical Congress 2021*, 2021. URL: https://elib.dlr.de/145177/.

[WMS+22] Armin Wedler, Marcus Gerhard Müller, Martin Schuster, Maximilian Durner, Peter Lehner, Andreas Dömel, Florian Steidle, Mallikarjuna Vayugundla, Ryo Sakagami, Lukas Meyer, Michal Smisek, Wolfgang Stürzl, Nicole Schmitz, Bernhard Vodermayer, Andre Fonseca Prince, Anouk Ehreiser, Emanuel Staudinger, Robert Pöhlmann, Siwei Zhang, Matthias Hellerer, Roy Lichtenheldt, Dennis Franke, Antoine Francois Xavier Pignede, Walter Schindler, Manuel Schütt, Bernhard Rebele, Wout Boerdijk, Riccardo Giubilato, Josef Reill, Moritz Kuhne, Jongseok Lee, Susanne Schröder, Sven Frohmann, Fabian Seel, Enrico Dietz, Rudolph Triebel, Neal Yi-Sheng Lii, Esther Bischoff, Christian Braun, Sean Kille, Kjetil Wormnes, Aaron Pereira, William Carey, A. P. Rossi, Laurenz Thomsen, Thorsten Graber, Thomas Krüger, Anko Börner, Patrick Irmisch, Kristin Bussmann, Gerhard Paar, Arnold Bauer, Stefan Völk, H. Rauer, Heinz-Wilhelm Hübers, Johann Bals, Sören Hohmann, Tamim Asfour, B. Foing, and Alin Olimpiu Albu-Schäffer. Finally! Insights into the ARCHESLunar Planetary Exploration Analogue Campaign on Etna in summer 2022. In IAC, editor, *73rd International Astronautical Congress, IAC 2022*, 2022. URL: https://elib.dlr.de/191389/.

Supervised Theses

[Abd19] Youssef Abdelhalem. Development of a Planning Algorithm for Multi-Robot Systems. bachelor thesis, Institute of Control Systems (IRS), Karlsruhe Institute of Technology (KIT), 2019.

[Dam19] Leonie Damaske. Erweiterung eines Branch-and-Price-Koordinationsansatzes zur Berücksichtigung komplexer Aufgaben in Multi-Roboter-Systemen. master thesis, Institute of Control Systems (IRS), Karlsruhe Institute of Technology (KIT), 2019.

[Guo21] Cheng Guo. Investigation of Reoptimization Approaches for the Coordination of Heterogeneous Robotic Teams. master thesis, Institute of Control Systems (IRS), Karlsruhe Institute of Technology (KIT), 2021.

[Hah22] Daniela Hahn. Entwicklung von Reoptimierungs-Methoden zur Koordination heterogener Multi-Roboter-Teams. master thesis, Institute of Control Systems (IRS), Karlsruhe Institute of Technology (KIT), 2022.

[Koh21] Saskia Kohn. Heuristische Reoptimierung für die Koordination heterogener Roboterteams. master thesis, Institute of Control Systems (IRS), Karlsruhe Institute of Technology (KIT), 2021.

[Mey19] Fabian Meyer. Entwicklung eines Planungsalgorithmus zur Sequenzierung und Zuweisung von Aufgaben in Multi-Roboter-Systemen. master thesis, Institute of Control Systems (IRS), Karlsruhe Institute of Technology (KIT), 2019.

[Sto20] Tim Stoll. Aufbau einer Simulationsumgebung für Multi-Roboter-Systeme. bachelor thesis, Institute of Control Systems (IRS), Karlsruhe Institute of Technology (KIT), 2020.

[Teu20] Jonas Teufel. Analysis of Existing Algorithms for the Coordination of Heterogeneous Robotic Teams. bachelor thesis, Institute of Control Systems (IRS), Karlsruhe Institute of Technology (KIT), 2020.

[Wö23] Robin Wöran. Entwicklung, Implementierung und Analyse von Ansätzen zur Berücksichtigung von komplexen Aufgaben in der optimalen Koordination von Multi-Roboter-Systemen. master thesis, Institute of Control Systems (IRS), Karlsruhe Institute of Technology (KIT), 2023.

Karlsruher Beiträge zur Regelungs- und Steuerungstechnik (ISSN 2511-6312)
Institut für Regelungs- und Steuerungssysteme

Band 01 Diehm, Gunter
Identifikation des menschlichen Bewegungsverhaltens auf der Basis von Primitiven.
ISBN 978-3-7315-0608-9

Band 02 Flad, Michael
Kooperative Regelungskonzepte auf Basis der Spieltheorie und deren Anwendung auf Fahrerassistenzsysteme.
ISBN 978-3-7315-0610-2

Band 03 Eckert, Marius
Modellbasierte Identifikation fraktionaler Systeme und ihre Anwendung auf die Lithium-Ionen-Zelle.
ISBN 978-3-7315-0690-4

Band 04 Krebs, Stefan
Intervallbeobachter für lineare parametervariante Systeme und deren Anwendung auf die Asynchronmaschine.
ISBN 978-3-7315-0857-1

Band 05 Kaspar, Stephan
Fahrdynamikuntersuchungen eines Elektrofahrzeugs mit Einzelrad-Hinterradantrieb.
ISBN 978-3-7315-0916-5

Band 06 Sauter, Patrick S.
Modellierung und zentrale prädiktive Regelung von multimodalen Energieverteilnetzen.
ISBN 978-3-7315-0963-9

Band 07 Kupper, Martin
Verteilte Zustandsschätzung fraktionaler Systeme und ihre Anwendung auf Lithium-Ionen-Batteriesysteme.
ISBN 978-3-7315-0971-4

Band 08 Merkert, Lennart
Optimal Scheduling of Combined Heat and Power Generation Considering Heating Grid Dynamics.
ISBN 978-3-7315-1056-7

Die Bände sind unter www.ksp.kit.edu als PDF frei verfügbar oder als Druckausgabe bestellbar.

Band 09 Ludwig, Julian
Automatisierte kooperative Transition einer Regelungsaufgabe zwischen Mensch und Maschine am Beispiel des hochautomatisierten Fahrens.
ISBN 978-3-7315-1069-7

Band 10 Inga Charaja, Juan Jairo
Inverse Dynamic Game Methods for Identification of Cooperative System Behavior.
ISBN 978-3-7315-1080-2

Band 11 Schnurr, Christoph Xaver
Ein Verfahren zur lexikographischen modellprädiktiven Regelung mit der Anwendung auf eine permanenterregte Synchronmaschine.
ISBN 978-3-7315-1095-6

Band 12 Schwab, Stefan
Guaranteed Verification of Dynamic Systems.
ISBN 978-3-7315-0965-3

Band 13 Pfeifer, Martin
Automated Model Generation and Observer Design for Interconnected Systems: A Port-Hamiltonian Approach.
ISBN 978-3-7315-1135-9

Band 14 König, Alexander
Absicherung hochautomatisierten Fahrens durch passiven virtuellen Dauerlauftest.
ISBN 978-3-7315-1141-0

Band 15 Stark, Oliver
Parameter- und Ordnungsidentifikation von fraktionalen Systemen mit einer Anwendung auf eine Lithium-Ionen-Batteriezelle.
ISBN 978-3-7315-1187-8

Band 16 Köpf, Florian
Adaptive Dynamic Programming: Solltrajektorienfolgeregelung und Konvergenzbedingungen.
ISBN 978-3-7315-1193-9

Band 17 Kölsch, Lukas
Dynamic Incentives for Optimal Control of Competitive Power Systems.
ISBN 978-3-7315-1209-7

Band 18 Schwartz, Manuel
Topologie-Optimierung eines radselektiv angesteuerten Fahrzeugs basierend auf einer optimalen Fahrzeugführungsregelung.
ISBN 978-3-7315-1222-6

Band 19 Rothfuß, Simon
Human-Machine Cooperative Decision Making.
ISBN 978-3-7315-1223-3

Band 20 Gellrich, Thomas Christoph
Dynamical Modeling and Control of Multiphase Heat Transport Systems Based on Loop Heat Pipes.
ISBN 978-3-7315-1231-8

Band 21 Maurer, Jona
Transactive Control of Coupled Electric Power and District Heating Networks.
ISBN 978-3-7315-1276-9

Band 22 Puccetti, Luca
Self-Learning Longitudinal Control for On-Road Vehicles.
ISBN 978-3-7315-1290-5

Band 23 Varga, Bálint
Limited Information Shared Control and its Applications to Large Vehicle Manipulators.
ISBN 978-3-7315-1325-4

Band 24 Walz, Fabian
Modellbasierte prädiktive Längsdynamikregelung für künftige Fahrerassistenz- und Automatisierungssysteme.
ISBN 978-3-7315-1331-5

Band 25 Strehle, Felix
A Framework for Decentralized Stabilization in Networked Energy System: A Passivity-Based Approach
ISBN 978-3-7315-1358-2

Band 26 Bischoff, Esther Sophie
Time-Extended Multi-Robot Task Allocation – A Reoptimization Framework with Provable Performance.
ISBN 978-3-7315-1449-7